COMPLEX WORLDS
Digital Culture, Rhetoric, and Professional Communication

Edited by

Adrienne P. Lamberti
University of Northern Iowa

and

Anne R. Richards
Kennesaw State University

Baywood's Technical Communications Series
Series Editor: CHARLES H. SIDES

Baywood Publishing Company, Inc.
AMITYVILLE, NEW YORK

Baywood Publishing Company, Inc.
26 Austin Avenue
P.O. Box 337
Amityville, NY 11701
(800) 638-7819
E-mail: baywood@baywood.com
Web site: baywood.com

Library of Congress Catalog Number: 2010004341
ISBN 978-0-89503-399-4 (cloth : alk. paper)
ISBN 978-0-89503-421-2 (epub)
ISBN 978-0-89503-422-9 (epdf)
http://dx.doi.org/10.2190/CWD

Library of Congress Cataloging-in-Publication Data

Complex worlds : digital culture, rhetoric, and professional communication / edited by Adrienne P. Lamberti and Anne R. Richards.
p. cm. -- (Baywood's technical communications series)
Includes bibliographical references and index.
ISBN 978-0-89503-399-4 (cloth : alk. paper) 1. Communication of technical information. 2. Internet literacy. 3. Digital media--Social aspects. 4. Telecommunication in education. 5. English language--Rhetoric. I. Lamberti, Adrienne Patrice. II. Richards, Anne R., 1961-
T10.5.C5925 2010
302.2--dc22

2010004341

Cover Photo by Iraj Omidvar, Schlossbrücke, Berlin 2009.

Table of Contents

Acknowledgments v

INTRODUCTION
Digital Divergence, Digital Complexity 1
Anne R. Richards and Adrienne P. Lamberti

PART I: TRANSFORMING ADVOCACY

CHAPTER 1
Cyberactivism, Viral Flash Activism, and Critical Literacy Pedagogy in the Age of *The Meatrix* 19
Eileen E. Schell

CHAPTER 2
Retracing the Footprints from Print to Digital: An Assessment of Textual Structure 35
Adrienne P. Lamberti

CHAPTER 3
The Fourth Estate in an Era of Digitally Mediated Democracy 57
Leonard Witt

PART II: SHAPING THE PROFESSIONS

CHAPTER 4
Gertrude Stein in QuickTime: Documenting Performance in the Digital Age 79
Jason Farman

CHAPTER 5
Digitizable Cultural Capital: Anticipations of Profit in the Web Market 95
John B. Killoran

CHAPTER 6
A Case Study of the Impact of Digital Documentation on Professional Change: The WPA Electronic Mailing List, Knowledge Network, and Community Outreach 117
Huiling Ding

PART III: BUILDING COMMUNITIES

CHAPTER 7
A South-North Online Collaboration between Professional Writing Students in Tunisia and the United States 133
Faiza Derbel and Anne R. Richards

CHAPTER 8
Meeting Online Friends Offline: A Comparison of South Korean and U.S. College Students' Differences in Self-Construal and Computer-Mediated Communication Preferences 157
Heeman Kim and William Faux

PART IV: INFORMING PEDAGOGY

CHAPTER 9
Teaching Effective Technology Use in Technical and Professional Communication Programs Based in Colleges of the Humanities 179
Laura McGrath

CHAPTER 10
Technical Communication Pedagogy and the Broadband Divide: Academic and Industrial Perspectives 195
Rudy McDaniel and Sherry Steward

CHAPTER 11
Sizing Up Single-Sourcing: Rhetorical Interventions for XML Documentation 213
Aimee Kendall Roundtree

Contributors 235
Index 239

Acknowledgments

For his belief in this project, we are profoundly grateful to Dr. Charles Sides, editor of Baywood's Technical Communications Series. We also thank our authors for their dedication, the participants in the review process for their patience, and the University of Northern Iowa and Kennesaw State University for their support. Finally, we thank our families, who have encouraged and strengthened us every step of the way.

INTRODUCTION

Digital Divergence, Digital Complexity

Anne R. Richards
and Adrienne P. Lamberti

When a respected scholarly journal published an article in 2006 stating that "college English should be new media" (Rice, 2006, p. 127), it unambiguously acknowledged the entrenchment of digital technology in the field of English studies. Testament to this altered landscape is the evolution in outlook evident in the National Council of Teachers of English's 1970 *Resolution on Media Literacy* and its 2005 *Summary Statement on Multimodal Literacies*. The *Resolution* is a two-paragraph call merely to "explore more vigorously the relationship of the learning and teaching of media literacy to other concerns of English instruction" (para. 2). More than 35 years later, the *Summary* offers numerous definitions and potentialities of digital technology for teachers and explores those "writing, design, and distribution processes that were formerly apportioned to other experts" (Declarations Concerning the Unique Capacities section, para. 15).

Especially noteworthy examples of the digital's influence in English studies are the growing gap between "students' view of technology and that of faculty," the movement of "digital expressions of scholarship . . . away from the standards of traditional peer-reviewed paper publication" (The New Media Consortium and EDUCAUSE Learning Initiative, 2007, p. 4), and the proliferation of new digitally mediated genres. Referencing Gunther Kress (2003) and Carolyn Miller (1984), for instance, S. Scott Graham and Brandon Whalen (2008) explore the potential for a new media form to incite several "generically distinct" social actions (p. 68). New media products, according to Graham and Whalen, are increasingly becoming "gestalt-shift genres," whose functions change during use (p. 89). Their example is a company e-card that, once opened and recognized as a holiday greeting, can be used as an interactive game.

We offer in this collection a series of essays mindful of digital technology's always complicated and often-fraught contexts, both in classrooms and in countless other communities of practice around the world. Our emphases on the situations of and audiences for digital discourse identify *Complex Worlds: Digital Culture, Rhetoric, and Professional Communication* as a rhetorically informed approach.

UNDERSTANDING DIGITAL DIVERGENCE

Throughout this collection, the term *digital divergence* is used to evoke the multifaceted and heterogeneous contexts, users, purposes, genres, and products of digital technology. This introduction frames the discussion and provides illustrations of digital divergence, first by exploring the topic as a tool of invention. No degree of facility with technology can illuminate the perplexing digital era we are all attempting to navigate. Thus, to facilitate a more penetrating awareness of digital culture, we offer a heuristic focused on the topic of digital technology. Table 1 provides examples of binary oppositions that help determine many professional communicators' ways of knowing. Each binary relation is followed by questions incorporating the issue of digital technology and disturbing essentialist assumptions about digitalization. A comprehensive list is not offered, but readers are encouraged to generate and to explore additional relations, such as agency/oppression, fluidity/stability, native/foreigner, traditional media/new media, linearity/ recursivity, profit motive/socially responsible investment, service/work, teacher/student, school/workplace, humanities/sciences, expert/learner, stranger/friend. We might explore, for instance, how liberation movements are employing digital surveillance and publishing technologies to broadcast oppressive government actions, reversing the effect of the Panopticon by anonymously training the gaze on their oppressors (privacy/surveillance).

This section concludes with extended discussions of concepts central to current scholarship and practice in professional communication: (a) the digital divide and (b) the user/designer relation, the exploration of which, we believe, can help unlatch key binaries at the very heart of our understanding of the world.

The Digital Divide

Scholars have tended to conceptualize the digital divide as an economically overdetermined phenomenon separating, in a broad stroke, rich from poor. What is more, the term denotes a complex of commonly held assumptions entailing binary oppositions such as East/West, agency/oppression, utopia/dystopia. Within the development community, this way of seeing has given rise to the questionable assumptions that the gap between rich and poor can be closed through diffusion of digital technology and that access in and of itself is a sufficient goal.

That the term "digital divide" has served as a rallying cry among multinational providers of telephony services is a situation now acknowledged and deconstructed by development practitioners (e.g., Deane, 2005). Not surprisingly, the academy has been implicated in the supply-side approach to the dissemination of digital technology. Mark Warschauer (2006), for example, observes in "Networking the Nile: Technology and Professional Development in Egypt" that economically powerful nations such as the United States are often eager purveyors of digital technology, offering scholarly participants perks that can prove nearly irresistible

Table 1. Potential Binaries for Consideration in Light of the Topos *Digital Technology,* and Example Questions

Culture/Nature	• How is the natural being integrated more fully into the lives of humans by means of digital technology? • How is digital technology being used to preserve natural resources?
Virtuality/Reality	• How is digital technology influencing human experience in the absence of computers? • How is virtual reality constrained by the real?
West/East; North/South; Developed/Developing, etc.	• To what extent does the West lack digital access, literacy, and empowerment? • To what extent does the East enjoy digital access, literacy, and empowerment?
Global/Local	• How does digital technology permit seepage of the global into the local (globalization)? • How does digital technology allow for seepage of the local into the global (localization)?
Metropole/Periphery	• How does digital technology allow for marginal populations to connect with, to identify with, and to become a part of the center? • How does digital technology allow for populations in the center to connect with, to identify with, and to become a part of the margins?
Work/Play	• How do digital work technologies increasingly enter into leisure time? • How do play and other forms of socializing for pleasure permeate the workplace by means of digital technology?
Individual/Group	• How does digital technology help integrate the individual into groups of various kinds? • How does digital technology undermine group dynamics?

(p. 164). Although digital diffusion is a pet project for many employees of industry, foreign-policy makers, and academics, who work together in mutually (though not always multilaterally) enriching collaborations, ill-conceived development programs focused on information communication technologies have, in James Deane's (2005) understated estimation, "wasted a lot of money" (p. 54). Ideally, programs to enhance access will take into account the possibilities of and impediments to digital technology and will incorporate the ongoing development of digital literacy and (where feasible) advocacy after access goals have been reached.

Additionally, whether economic class differences, which influence the quality of *all* areas of life—from choice of footwear to transportation to communication medium—are especially grave with regard to the digital simply has not been established (Souter, 2005, p. 14). A number of prominent figures in the field of development are asserting that there is "no substantive consensus on the value, role, or significance" (p. 11) of digital technologies in helping to bridge the world's economic chasms (e.g., Deane, 2005). For some critics, the very idea of a digital divide is a "red herring," distracting development practitioners from the facts that their "industry is notorious for going through fads and fashions" (p. 60) and that the United Nations Millennium Development Goals do not recognize digital access.

In the field of professional communication, scholars have tended to assume that digital access, literacy, and advocacy are linked primarily to economic status. And to be sure, the economic component of digital divergence must not be minimized. Data from 2003 indicated that 85% of the world's users of Internet technology were from comparatively affluent nations, although these users accounted for only 22% of the world's population (Hudson, 2005, p. 43).

Since "possession of high income, and occupational and educational resources gives access to additional information, which may further exacerbate existing social disparities and strengthen advantage" (Willis & Tranter, 2006, p. 45) in a process known as *homophily*, digital adoption reflects the fact, or perhaps ensures, that to a certain (and at this writing it seems to a great) extent, the world's rich are getting richer and its poor poorer. What is more, although many Africans have obtained access to mobile phone technology in the last decade, this advantage at times has appeared to constitute a long-term impediment, for the impetus to lay the copper-based fixed lines that have been required for both land-line phone and Internet access has been dulled. Likely for this reason, in 2003, "the price of Internet access was higher in Africa than anywhere else in the world, and completely out of reach of most Africans" (Hudson, 2005, p. 43).

If we look, however, at class difference in terms of one key communications technology—the telephone—we find that this gap has decreased very significantly over the last few decades. In 1982, high-income or upper-middle-income households owned 96% of phones around the world; in 2002, that percentage had fallen to 61% (Kelly, 2005, p. 23). And in 2008, mobile phones were "within reach of 60–70%" of Africa's population (Southwood, 2008).

These figures suggest that Everett M. Rogers' (1962) innovation-adoption curve is relevant to the adoption of telephone technologies, as it clearly is for television, which is now ubiquitous across the planet, from the virtual palaces of Seattle's silicon elite to the humblest troglodyte homes of the Atlas Mountains. Likewise, noting that "the process of 'catch-up' is occurring much more rapidly among newer services than for older ones" (p. 29), Tim Kelly (2005) observes that

> Because technological change is also accelerating, and [information and communication technology (ICT)] innovations are being introduced on an almost daily basis, the popular impression is that the digital divide is expanding, because each succeeding ICT innovation starts the diffusion process all over

> again. Innovations tend to start in the richer countries and spread to the poorer ones. . . . As the evidence shows [however], for the main bearer networks the digital divide is narrowing as diffusion spreads, and in most cases at an accelerating pace. (p. 27)

Supporting his observation is the fact that Internet and mobile phone usage in economically developing lands was, in 2003, approximately the same as it had been in economically developed lands 5 years before (Hudson, 2005). Over the period 2002–2008, Internet usage grew exponentially in the Middle East, Africa, and Latin America/the Caribbean, increasing 1100%, 1296%, and 860%, respectively. In contrast, in the comparatively saturated markets of North America and Europe, Internet usage increased only 132% and 275%, respectively (Internet World Stats, 2009). According to Russell Southwood (2008), in national surveys from 12 African countries of varying per capita incomes, "between 2–15% of the population use the Internet (except in the two poorest countries) and 1–8% use it on a daily basis (except for the four poorest countries). On this basis, there might easily be tens of thousands or hundreds of thousands of [African] broadband subscribers depending on the size of country" (para. 11). We should bear in mind that ubiquity of digital technology such as the Internet is being accelerated in part through shifts in design that increasingly fold multiple products into one. Whereas the rotary phone had a very limited range of functions, cellular phones offer options such as Internet service, contact presence information, computer operating systems, GPS and gaming capabilities, connectivity to corporate intranets and databases, and remote access to printers and fax machines.

In consideration of these data, we are in agreement with fellow researchers who find that any "simplistic notion of a digital divide is no longer useful" (Willis & Tranter, 2006, p. 56). According to Suzanne Willis and Bruce Tranter (2006), the reality of Internet usage, as studied across the disciplines, simply does "not resemble the 'digital divide' as it is often conceptualized" (p. 54). Many scholars insist that the concept of a divide should be exchanged for that of a web of technology-use situations diverging in light of multiple factors such as wealth, traditional literacy, age (e.g., Deane, 2005), geography (e.g., Kelly, 2005), human-resource management (e.g., Hinson, 2006), professional status (e.g., Willis & Tranter, 2006), gender (e.g., Subramony, 2007), race (e.g., Tolbert, Mossberger, & Stansbury, 2003), teacher training (e.g., Ololube, 2006), research and development investment (e.g., Norris, 2001), foreign investment (e.g., Robison & Crenshaw, 2002), and degree of democratization (e.g., Elijah & Ogunlade, 2006).

By insisting on the term *digital divergence* throughout this collection, we mean, among other things, to indicate our distance from the term *digital divide,* which clearly is being critiqued vigorously and increasingly is being recast as a web of complex, technologically mediated material and discursive situations in which digital access, literacy, and advocacy are at stake. We hope that by reframing these issues in terms of divergence we can offer members of our discipline a way of more faithfully understanding technology use in our own local situations and around the world.

We should note, however, that by referring to divergence in the context of digital culture, we are neither advocating nor engaging in the popularization of a scientific concept. For colleagues in the sciences, divergent systems are those that become progressively more different over time; convergent systems become progressively more of a piece, more homogenous (e.g., Phillips, 2004). We cannot call digital culture divergent in this sense because, as we have already acknowledged, the process of homophily may be stabilizing and strengthening structural inequalities based in digital differences. We also must remind ourselves that as we explore digital culture and come face to face with its seeming unpredictability, we may be discovering differences that have preexisted our explorations. (For instance, we may come to recognize the counterintuitive notion that there are many wealthy people—men, citizens of powerful democracies—who are not empowered through digital culture, while many impoverished people—women, citizens of authoritarian regimes—are.) In short, whether digital culture in any veritably scientific sense is diverging (evolving toward greater difference) or converging (evolving toward greater sameness) is not our concern. But strategic use of the term *divergence* in lieu of *difference* or *divide* allows us to highlight the fact that the panoply of economic, political, and social systems being transformed in the era of globalization are likely to have a great deal in common and to be *similar up to a point.* For as Doreen Starke-Meyerring, Ann Hill Duin, and Talene Palvetzian (2007) remind us, no matter the framework in which globalization is understood, its "processes have a concrete material basis: They are characterized by a far-reaching shift in the means of production," namely, "the use of the Internet for the globally distributed production and delivery of services" (p. 141).

Users/Designers

As we have just explained, the term *digital divide* suggests an intractable opposition between digital culture's "haves" and "have nots." Yet another binary that the topic *digital technology* can help us rethink is that of user/designer, two identities that in the last century existed in relative juxtaposition as a result of the dominance of process engineering and mass production.

The computer gaming community is perhaps unique in terms of the ongoing design efforts of its users. Users create an array of discourse around the game, including communication in chat rooms and on bulletin boards, guild Web sites, clan forums, and blogs. But these are by no means the only genres or media that users generate: T. L. Taylor also identifies art sites and "fan produced movies" (2006, p. 57), comics (p. 17), "fan art and stories, and even game modifications that contribute to players feeling a bond and connection" (p. 57) to their server and to one another, "game guides, walk-throughs, answers to frequently asked questions (FAQs), maps, object and monster databases, third-party message boards and mailing lists, play norms, server guidelines, modifications, plug-ins, strategies and strategy guides, auctions/trading, tweaks to user interfaces (UI), macro sharing . . . counter-narratives" (p. 155).

Gamers are involved in a full-blown collaboration not only with other gamers but with game designers, and constantly push, pull, tweak, transform, the games they play. "More radically put," writes Taylor, "the very product of the game is not constructed simply by the designers or publisher, nor contained with the boxed product, but produced only in conjunction with the players" (p. 126). Many, if not most, cutting-edge developments in gaming can be traced to "power players," who manage through their constant intimate association with a product to localize and to exploit it in ways that far outstrip the expectations of its officially recognized designers.

Also interested in the impact of users on product design, Huatong Sun (2006) theorizes the "messiness" attending digital technology (p. 460). By studying the interactions among localization design efforts, IT products and services, and user responses in order to surface the "complexities and local contexts" of text messaging, Sun argues that current theory tends to compartmentalize technological embodiment, context, practice, and communication. In response, she identifies the "where, when, who, how, and what" influencing a U.S. and a Chinese cell phone user's texting activities (p. 467), articulating a "use story" for their daily lives (p. 473). Sun demonstrates how users attempt to localize a product in an effort to work around technological challenges and when successful, "rescue" weak localization efforts made at the design phase. She suggests that these rescues present opportunities for designers to develop more supportive localization efforts (p. 475).

EXPLORATIONS IN DIGITAL COMPLEXITY: KEY AREAS OF INTEREST

As Byron Hawk (2007) notes in his discussion of vitalism and the field of rhetoric and composition, a "new paradigm built around complexity could produce a post-dialectical understanding of contemporary pedagogies of [writing] invention for the emerging scene produced by digital technology" (p. 7). So far, we have explored such a postdialectical understanding. We next explore what we believe are and will remain key topics in the rhetorical study of digital culture.

Public and Corporate Governance

Langdon Winner (2003, 2005) argues that digital technology has the potential to strengthen the integrity of decentralized political systems. Indeed, digital culture and democracy have the potential to enhance or to undermine each other, as Foucault's insights into the mutual constitution of systems of governance and instruments of technical rationality suggest. But Winner warns that the "euphoria" following the emergence of a new technology can preclude "serious reflection about whether the arrival of a new technological device or system will truly be beneficial to political freedom and democratic governance" (p. 124). Rather than assume that technology is neutral or always a form of progress and thus to allow other actors such as the private sector to determine how a technology will be used, Winner argues that it has the greatest potential to enhance culture when discussed by a chorus of participants, public and private (p. 133). Public participation in debates surrounding the use of digital technology is vibrant in many regions of the world and often exists despite the disapproving eye—and disciplining hand—of government.

Some transnational corporations have budgets that exceed the gross domestic products of certain nations. A response to the growing power of these companies is socially responsible investing, which increased from 40 million to 2.16 trillion dollars in the period 1984–2003 (Social Investment Forum, 2003, p. 2; cited in Starke-Meyerring et al., 2007). What is more, stakeholder-rights movements are increasingly striving to influence "corporate practices regarding labor rights, health, diversity, and environmental protection as well as . . . crossover resolutions to link social and environmental performance to corporate governance, including executive pay" (p. 144). Because of the global reach of transnational corporations, related investing and stakeholder-rights movements rely on digital technology to grow membership and to publicize their causes.

The Globalizing Academy

Laura Gurak and Ann Hill Duin (2004) acknowledge fears among some academics that, in its influence on pedagogy and epistemology, digital technology may be compromising the very foundations of the academy. Although they insist that new technologies, by augmenting scholarly networks and facilitating research, can help members of the academy adhere to the tenets of "our traditional institutions, with an emphasis on research-based teaching" (p. 197), they observe that to resist or to reject digital technology increases the risk that institutions of learning will be hijacked by technologically savvy "for-profit universities and learning centers, where students will gain a diploma but lose contact with the long history and tradition" of the academy (p. 197). In contrast, Barbara Mirel and Nicholas Johnson (2006) posit that this history and tradition has the potential to be transformed positively by digital collaborations such as Internet2, "a high-performance, high-speed, next-generation Internet designed for the needs of remote scientific collaborations, distance education, and health care" (p. 329). Currently "more than 200 research universities, various government partners, and corporate sponsors" (p. 331) work together under the auspices of this consortium.

Complicating the mix of business and academic interests is the element of international partnerships. Jennie Dautermann (2005) suggests that the Fulbright and Peace Corps programs are, with a new emphasis, currently recruiting academics with expert knowledge in business and technology. Likewise, she cites anecdotal evidence that there is an increasing interest abroad in employing U.S. faculty in English to teach business and technical writing (p. 149). Indeed, Philip Altbach (2004b, p. 22; cited in Starke-Meyerring et al., 2007) asserts that "we are at the beginning of the era of transnational higher education, in which academic institutions from one country operate in another, academic programs are jointly offered by universities from different countries, and higher education is delivered through distance technologies."

Professional Communication Pedagogy

Teachers in all disciplines encounter pressures to teach online, often as a result of funding cutbacks and/or rising enrollments. Regrettably, teachers in the humanities may be especially unprepared for the shift to digitally mediated instruction. Studying the training of online teachers in technical communication, Kelli Cargile Cook (2007) enumerates the applications that online teachers ideally will be skilled in using: "Course management tools that promote interaction, such as threaded discussions or forums, file-sharing and e-mail utilities, and chat applications. They may also include multimedia components such as audio, video, and other websites" (p. 58). In a recent online graduate class in which Cook taught 18 students, her course archive ultimately included more than 3,000 entries in various online forums (p. 63). Given the daunting amount of communication for which online teachers are responsible, they also need to learn how to employ the advanced tools offered with course management softwares such as WebCT or Blackboard. Additionally, teachers will adjust their own discourse to become "shardlike"; that is, they will begin composing texts that are more concise and pointed, and thus likely to be more useful when scattered across the course interface. In Cook's view, the course archive ideally will be an outcome not of *exploratory* hypertext created for students to navigate according to a virtual road map provided by the teacher, but of *constructive* hypertext, which teacher and students co-create. Such an archive will be rhizomatic, in that it is "multiple and changeable, with threads or roots that can emerge randomly at indefinite times and places" (p. 75).

Melinda Turnley (2007) argues for the power of critical approaches to technology when they are used to enrich the learning experience of students engaging in service writing in the digital age. By asking students to work with local nonprofit organizations and to "adopt a critical, situated approach" toward their use of technologies there, she found that students began "to view technology as enmeshed within their clients' larger organizational cultures and discursive formations" (p. 116). Outcomes included a change in focus, away from specific hardware and software, toward an enhanced understanding of technology's influence on worker relationships; institutions themselves; their discursive, procedural, ethical, ideological, and policy characteristics; and individual workers.

The types of institutions with which students might work as service-learning interns are likely to be wired, and their members are likely to appreciate competence in the rhetorical understanding and practical application of digital technologies. For instance, volunteers with Hands On Network, which recognizes the differences among various populations' abilities to access service organizations, have identified the most productive ways in which nonprofits can use electronic and printed calendars to recruit and to update volunteers (Using a project calendar to assist with recruiting and scheduling volunteers, n.d., Action section, para. 3–6). And an employee of telecommunications megacorporation Qwest (Kearns, 2001) has advocated that the best practice in alleviating unequal access to digital technology lies in the creation of partnerships in the form of "mutual, supportive arrangement[s] between organizations, agencies, businesses, and/or communities with the purpose of addressing a problematic situation" (What Is a Partnership? section, para. 1). Such partnerships can help individuals and institutions surmount "huge obstacles [to access] such as our communications infrastructure, our education system, and our values as a society" (How Can Partnerships . . . section, para. 2).

Writing in the Professions

According to Starke-Meyerring and colleagues, the spread of distributed work across global networks, the importance of digital technologies to users and others, the proliferation of policies at all levels regarding the workplace, and public opinion about the evolution of the role of transnational corporations in the political sphere and elsewhere (2007, pp. 141–142) are transforming careers across industry.

Yet Jonathan Adams, Margaret DeFleur, and Gary Heald (2007) observe that digital literacies remain suspect in certain professions. As noted, academics in traditional educational settings can be suspicious of for-profit, online universities. This attitude is shared by many employers, including the health care administrators who took part in Adams and colleagues' study and were found to "overwhelmingly" prefer job applicants with traditional credentials over applicants "holding the degree earned online or partly online" (p. 292).

The American Society of Magazine Editors (n.d.) embraces a rhetorical stance toward digital literacy, advising its members that "credibility is key to the success of all digital-media businesses with an editorial component. Users must trust the advice and information given, just as they do that of offline brands" (para. 1). The society consequently recommends that, in the construction of magazine Web sites, conventions be adopted that are "subject to change as the medium evolves" (para. 2). Jeffrey Penka (2003) describes how an innovative digital reference system called QuestionPoint has inspired library staff and others in information technology fields to develop new digitally informed practices for locating, using, amassing, and maintaining resources and to reexamine how a digital service affects the quality of service to library patrons (Technical Standards section, para. 3). Another practical result can be seen in the Washington State Library's *Digital Best Practices* site, which demonstrates how a display in digital form should be mounted in light of "Project Management," "Collection [selection]," "Technology [requirements]," and "Funding" considerations distinct from those shaping physical, glass-case displays (Welcome to the Washington State Library Digital Best Practices, n.d., Planning a digital project section).

Ruth Bolton (2003) observes that in marketing and public relations, "marketers have only begun to address" the benefits suggested by digitally delivered advertising, in part because of perceived irreconcilable differences between the Internet and traditional marketing media such as telephone and television (p. 44). Bolton encourages an approach to design that blends the conventional electronic and the contemporary digital in an attempt to reach a broader range of consumers (p. 43). Such hybridized consumer advertising would take on added dimensions if played out in a technologically transformed economic context such as that described by Shoshana Zuboff and James Maxmin (2004). Their chronicle of digital technology's economic impact heralds the decline of mass production and consumption in favor of a "support" economy, customizing the production and delivery of goods according to individuals' unique needs (p. 7).

CHAPTERS IN THIS COLLECTION

Conceptualized, written, and edited for a wide range of readers interested in digital culture, this anthology offers a broad focus on the contexts of professional communication and its effective teaching and practice. In their appraisals of the technological issues currently transforming not just professional communication but related fields such as journalism, rhetoric, and English-language teaching, the chapters in *Complex Worlds: Digital Culture, Rhetoric, and Professional Communication* present a methodological pluralism. Grappling with the discursive and material shifts occasioned by new technology in educational contexts and beyond, *Complex Worlds* studies professional communication in light of the key concepts of digital access, literacy, and advocacy. Several chapters approach the economic structures influencing access to digital technology as an a priori and intractable reality; others treat such an assumption as unnecessarily dire. All discussions, however, provide nuanced constructions of digital divergence in light of its expression in industrial and/or educational practices.

The first part of *Complex Worlds*, "Transforming Advocacy," examines the intersections between digital technology and agency in computer-mediated participatory cultures. Eileen E. Schell's "Cyberactivism, Viral Flash Activism, and Critical Literacy Pedagogy in the Age of *The Meatrix*" describes the growth of cyberactivism by means of viral Flash films. Such cyberactivist efforts, Schell explains, can be studied in an effort to enhance critical literacy skills. Adrienne P. Lamberti's "Retracing the Footprints from Print to Digital: An Assessment of Textual Structure" explores why the prospect of textual restructuring as brought on by digital forms has elicited such strong scholarly responses. These evince evolving conceptualizations of authorship, understandings of reader analysis and comprehension, and forms of advocacy for multivocal inclusivity. "The Fourth Estate in an Era of Digitally Mediated Democracy," by Leonard Witt, examines the effects that epochal changes in news delivery are having on the production of credible news and the role that online citizen journalists are playing in overthrowing the traditional journalistic paradigm. A key effect of this revolution is the breakdown of long-cherished distinctions between consumers and producers of news.

The second part of the anthology, "Shaping the Professions," examines ways in which digital technology is influencing the work lives of communication practitioners. In "Gertrude Stein in QuickTime: Documenting Performance in the Digital Age," Jason Farman chronicles the creation of an interactive CD-ROM by a New York performance troupe. The composition process that Farman relates involves an intriguing mix of genres and media, for a play by Gertrude Stein is rewritten as a second play, which is itself recorded on a CD-ROM and finally is translated into an interactive DVD. John B. Killoran's "Digitizable Cultural Capital: Anticipations of Profit in the Web Market" applies Pierre Bourdieu's criticism of academic expectations regarding portfolios—specifically the new genre of the digitized résumé—and finds that these expectations can negatively affect the prospects of students who lack the cultural capital necessary to master that genre. In "A Case Study of the Impact of Digital Documentation on Professional Change: The WPA Electronic Mailing List, Knowledge Network, and Community Outreach," Huiling Ding explores how the Writing Program Administrators' electronic mailing list assists them in articulating their work as not "merely" service but influential acts of professional outreach. Illustrating how this digitally mediated communication mode can facilitate knowledge sharing among practitioners with differing levels of professional expertise and resources, her research reveals differences in administrators' understandings of what constitutes professionalism.

The third part of *Complex Worlds*, "Building Communities," concerns the remarkable potential of digital technology to deepen our understanding of others and to create communities that expand our social lives. In "A South-North Online Collaboration between Professional Writing Students in Tunisia and the United States," Faiza Derbel and Anne R. Richards recount a digitally mediated collaborative project between U.S. and Tunisian classrooms with divergent technological capabilities. The authors suggest ways in which digital e-collaborations between classrooms of very different economic means and cultural outlooks can be mutually beneficial. In "Meeting Online Friends Offline: A Comparison of South Korean and U.S. College Students' Differences in Self-Construal and Computer-Mediated Communication Preferences," Heeman Kim and William Faux conduct a cross-cultural exploration of computer-mediated communication utilities, uncovering surprising differences and similarities between U.S. and South Korean users of new media and calling into question the lingering Western stereotype of Asians as passive communicators.

The final part of *Complex Worlds,* "Informing Pedagogy," offers recommendations relevant to digitally informed communication classrooms. In "Teaching Effective Technology Use in Technical and Professional Communication Programs Based in Colleges of the Humanities," Laura McGrath offers pedagogical rationales and strategies for helping students cultivate digital literacy skills, with the goal of enabling meaningful use of technology in a wide range of workplace situations. Referencing key concerns discussed in the chapters of this section, McGrath resolves that "an integrated, rhetoric-based approach to technology, combined with outward-reaching collaborative and strategic partnerships" can mitigate tensions over digital technology that are often experienced by teachers of professional communication. Rudy McDaniel and Sherry Steward's "Technical Communication Pedagogy and the Broadband Divide: Academic and Industrial Perspectives" explores how the increased availability of broadband Internet service presents academics working in the field of technical communication and practitioners outside of academe the opportunity to leverage the "petrol of the new global economy" (Southwood, 2008). These authors argue for increased attention to the technological component of rhetorically focused technical communication pedagogy, which in their view may not be preparing students for the workplace. Aimee Kendall Roundtree's "Sizing Up Single-Sourcing: Rhetorical Interventions for XML Documentation" explains why professional communicators coding in this extensible language must not forget their lessons in rhetoric. Invoking such lessons in the workplace is crucial, Kendall Roundtree argues, if professional communicators are to create documents that truly serve their target audiences.

In the communities we enter daily, we increasingly encounter new and often perplexing technological scenarios. At the same time, the digital has the potential to help us generate clearer and more sophisticated understandings of our worlds—global and local, material and imaginative—in ways heretofore impossible. *Complex Worlds: Digital Culture, Rhetoric, and Professional Communication* offers teachers and students of professional communication a collection of provocative responses to some of the most urgent challenges facing communicators today.

REFERENCES

Adams, J., DeFleur, M. H., & Heald, G. R. (2007). The acceptability of credentials earned online for obtaining employment in the health care professions. *Communication Education, 56*(3), 292–307.

American Society of Magazine Editors. (n.d.). *Best practices for digital media.* Retrieved July 16, 2007, from http://www.magazine.org/Editorial/Guidelines/Best_Practices_for_Digital_Media

Bolton, R. N. (2003). Marketing challenges of e-services. *Communications of the ACM, 46*(6), 43–44.

Cook, K. C. (2007). Immersion in a digital pool: Training prospective online instructors in online environments. *Technical Communication Quarterly, 16*(1), 55–82.

Dautermann, J. (2005). Teaching business and technical writing in China: Confronting assumptions and practices in home practices. *Technical Communication Quarterly, 14*(2), 141–159.

Deane, J. (2005). Not a telecoms, nor a digital, but an information divide. In G. Milward-Oliver (Ed.), *Maitland + 20: Fixing the missing link* (pp. 51–64). Bradford on Avon, UK: The Anima Centre Limited.

Elijah, O. A., & Ogunlade, I. (2006). Analysis of the uses of information and communication technology for gender empowerment and sustainable poverty alleviation in Nigeria. *International Journal of Education and Development Using Information and Communication Technology, 2*(3), 45–80.

Graham, S. S., & Whalen, B. (2008). Mode, medium, and genre: A case study of decisions in new-media design. *Journal of Business and Technical Communication, 22*(1), 65–91.

Gurak, L., & Duin, A. H. (2004). The impact of the Internet and digital technologies on teaching and research in technical communication. *Technical Communication Quarterly, 13*(2), 187–198.

Hawk, B. (2007). *A counter-history of composition: Toward methodologies of complexity.* Pittsburgh, PA: University of Pittsburgh Press.

Hinson, R. (2006). Internet adoption amongst final year students in Ghana's oldest business schools. *Library Review, 55*(5), 314–323.

Hudson, H. E. (2005). From missing links to digital divides: Progress and change since the *Maitland Report*. In G. Milward-Oliver (Ed.), *Maitland + 20: Fixing the missing link* (pp. 35–50). Bradford on Avon, UK: The Anima Centre Limited.

Internet World Stats. (2009). *Internet usage statistics: The Internet big picture.* Retrieved May 25, 2009, from http://www.internetworldstats.com/stats.htm

Kearns, T. (2001). *Using partnerships to bridge the digital divide within the disability community.* Retrieved July 18, 2007, from http://www.icdri.org/DD/using_partnerships_to_bridge_the.htm

Kelly, T. (2005). Twenty years of measuring the missing link. In G. Milward-Oliver (Ed.), *Maitland + 20: Fixing the missing link* (pp. 21–34). Bradford on Avon, UK: The Anima Centre Limited.

Mirel, B., & Johnson, N. (2006). Social determinants of preparing a cyber-infrastructure innovation for diffusion. *Technical Communication Quarterly, 15*(3), 329–353.

National Council of Teachers of English. (1970). *Resolution on media literacy.* Retrieved July 11, 2007, from http://www.ncte.org/about/over/positions/category/media/107522.htm

National Council of Teachers of English. (2005). *Summary statement on multimodal literacies.* Retrieved July 11, 2007, from http://www.ncte.org/about/over/positions/category/media/123213.htm

The New Media Consortium and EDUCAUSE Learning Initiative. (2007). *The horizon report.* Retrieved July 11, 2007, from http://www.nmc.org/horizon/2007/report

Norris, P. (2001). *Digital divide: Civic engagement, information poverty, and the Internet worldwide.* Cambridge, UK: Cambridge University Press.

Ololube, N. P. (2006). Appraising the relationship between ICT usage and integration and the standard of teacher education programs in a developing country. *International Journal of Education and Developing Using Information and Communication Technology, 2*(3), 70–86.

Penka, J. T. (2003). *The technological challenges of digital reference: An overview.* Retrieved July 14, 2007, from http://www.dlib.org/dlib/february03/penka/02penka.html

Phillips, J. D. (2004). Divergence, sensitivity, and nonequilibrium in ecosystems. *Geographical Analysis, 36*(4), 369–383.

Rice, J. (2006). Networks and new media. *College English, 69*(2), 127–133.

Robison, K. K., & Crenshaw, E. M. (2002). Post-industrial transformations and cyber-space: A cross-national analysis of Internet development. *Social Science Research, 31*, 334–363.

Rogers, E. M. (1962). *Diffusion of innovation*. New York: Free Press.

Salvo, M. J. (2002). Critical engagement with technology in the computer classroom. *Technical Communication Quarterly, 11*(3), 317–338.

Souter, D. (2005). Then and now: What would be the remit of a modern-day Maitland Commission? In G. Milward-Oliver (Ed.), *Maitland + 20: Fixing the missing link* (pp. 3–20). Bradford on Avon, UK: The Anima Centre Limited.

Southwood, R. (2008). *Trends in technology. Global Information Society Watch 2008.* Retrieved May 25, 2009, from http://www.giswatch.org/gisw2008/thematic/TrendsInTechnology.html

Starke-Meyerring, D., Duin, A. H., & Palvetzian, T. (2007). Global partnerships: Positioning technical communication programs in the context of globalization. *Technical Communication Quarterly, 16*(2), 139–174.

Subramony, D. P. (2007). Understanding the complex dimensions of the digital divide: Lessons learned in the Alaskan arctic. *The Journal of Negro Education, 76*(1), 57–67.

Sun, H. (2006). The triumph of users: Achieving cultural usability goals with user localization. *Technical Communication Quarterly, 15*(4), 457–481.

Taylor, T. L. (2006). *Play between worlds: Exploring online game culture.* Cambridge, MA: MIT Press.

Tolbert, C. J., Mossberger, K., & Stansbury, M. (2003). *Virtual inequality: Beyond the digital divide.* Washington, DC: Georgetown University Press.

Turnley, M. (2007). Integrating critical approaches to technology and service-learning projects. *Technical Communication Quarterly, 16*(1), 103–123.

Using a project calendar to assist with recruiting and scheduling volunteers. (n.d.). Retrieved July 14, 2007, from http://nationalserviceresources.org/epicenter/practices/index.php?ep_action= view&ep_id=4828

Warschauer, M. (2006). Networking the Nile: Technology and professional development in Egypt. In J. A. Inman & B. L. Hewett (Eds.), *Technology and English studies: Innovative professional paths* (pp. 163–173). Mahwah, NJ: Lawrence Erlbaum Associates.

Welcome to the Washington State Library digital best practices. (n.d.). Retrieved July 16, 2007, from http://digitalwa.statelib.wa.gov/newsite/best.htm

Willis, S., & Tranter, B. (2006). Beyond the "digital divide": Internet diffusion and inequality in Australia. *Journal of Sociology, 42*(1), 43–59.

Winner, L. (2003). The Internet and dreams of democratic renewal. In D. Anderson & M. Cornfield (Eds.), *The civic Web: Online politics and democratic values* (pp. 167–183). Lanham, MD: Rowman and Littlefield.

Winner, L. (2005). Technological euphoria and contemporary citizenship. *Techne: Journal of the Society for Philosophy and Technology, 9*(1), 124–133.

Zuboff, S., & Maxmin, J. (2004). *The support economy: Why corporations are railing individuals and the next episode of capitalism*. New York: Penguin.

PART I

Transforming Advocacy

http://dx.doi.org/10.2190/CWDC1

CHAPTER 1

Cyberactivism, Viral Flash Activism, and Critical Literacy Pedagogy in the Age of *The Meatrix*

Eileen E. Schell

OVERVIEW

This chapter examines a specific example of cyberactivism—the viral Flash activist film *The Meatrix*—to illuminate not just this technology's possibilities for but also its troubling uses in digital organizing and digital information dissemination. *The Meatrix* offers a critical literacy problem-posing exercise in exploring the kinds of "oppositional consciousness" and "alternative solidarities" (Bousquet, 2003, p. vii) that are necessary if professional communication students are to become mindful users of the products of digital technology.

This chapter examines a specific example of cyberactivism, the viral Flash activist film *The Meatrix*, which offers for professional communication classrooms a heuristic for discussions that may benefit a wide range of rural, urban, and suburban communities and communicators in their discussions of social topics such as factory farming. Whether they realize it or not, most U.S. residents buy milk, eggs, and meat that have been factory farmed, and *The Meatrix* can prompt students to think of factory farms not as a problem "out there" (out of sight, out of mind), but as a problem for everyone. This chapter analyzes *The Meatrix* as a case study of the possibilities and limitations of digital organizing and digital information dissemination, as a critical literacy problem-posing exercise, and as an opportunity for exploring the kinds of "oppositional consciousness" and "alternative solidarities" (Bousquet, 2003, p. vii) that are necessary if professional communication students are to become mindful users of the products of digital technology.

Although it encourages solidarity building among rural, urban, and suburban communities around the issue of food consumption, the film also presents factory farming as an inevitable logic thrust upon us by nameless and faceless agricultural corporations. The film therefore ignores how such an "innovation," whose stakeholders include faculty and administrators employed by land-grant institutions, has evolved as part of the progress-and-efficiency narrative generated by late capitalism. As Robert P. Martin (2003), Executive Director of the Pew Commission on Industrial Farm Animal Production, notes, the meat industry has a significant influence over "academic research, agricultural policy development, government regulation, and enforcement" (p. viii). In fact, Martin warns that the "agro-industrial complex—an alliance of agriculture commodity groups, scientists at academic institutions who are paid by industry, and Capitol Hill—is a concern in animal food production in the 21st century" (p. viii). In the instance of factory farming, commercial interests (e.g., the meat industry) and public interests (e.g., land-grant institutions) are closely intertwined.

My writing pedagogy on the issue of factory farming is inspired by personal concerns about the agro-industrial complex that are reflected in Derek Owens' (2000) introduction of questions of sustainability in *Composition and Sustainability*, as well as in Robert Brooke's (2003) work on place-conscious education. Citing Haas and Nachtigal (1998), Brooke urges us to consider place-conscious education—an understanding of one's location, responsibility to place and to community, and civic life—as a significant part of a literacy curriculum, whether in K–12 or in college (pp. 12–13). And today the issue of place is a standard consideration in professional communication literature and practice. Scholars such as Joann Hackos (1994) and Mahesh S. Raisinghani (2007) long have focused on professional writing efforts at "localization, a term used these days to signify the differences a text must undergo to be effectively exported to a new site" (Killingsworth, 2005, p. 360). More recent characterizations of the concept of location, however, have experienced a shift to more closely approximate the relation between place and responsibility. M. Jimmie Killingsworth's (2005) revisiting of location, for instance, reconceptualizes the topic as a different kind of place consciousness and further as a pedagogical heuristic that can energize professional communication curricula. Killingsworth looks to ecopoetic theory, and in particular to the metaphors one invokes when thinking of a specific locale, to investigate what happens when a writer sees "the text of the document and the discourse community . . . [as a] substitute for face-to-face interaction" (p. 368). "Ecopoetically informed" professional writing instruction, he argues, would emphasize or "restore" the faces, the earth, the heartbeats that are lost when text is viewed as an indifferent placeholder for personal contact, and in doing so would reorient writers to their documents' implication in the human condition (p. 369).

Bringing place-conscious education and questions of sustainability to the professional communication classroom gives us an opportunity to think about the simultaneously simple and complex issue of our tables and what they contain: Where, in fact, does our food come from? What do we know and what do we not know about how food arrives in grocery stores and distribution outlets? Is there a hidden food-literacy curriculum, and if so how can we become critically literate about it? By extension, how can cyberactivism shed light on the public's food literacies? By exploring these questions in the section that follows, I argue for the need to understand the limits and potentials of digital activism and lay the groundwork for a consideration of the *The Meatrix* as an example of cyberactivism and specifically of viral Flash activism.

CYBERACTIVISM AND DIGITAL DIVERGENCE

Activists and nonprofit volunteers who have engaged in social change throughout the 1980s and 1990s and into the millennium can testify firsthand to the possibilities of organizing via online activism, or *cyberactivism*. Using e-mail, electronic mailing lists, chat rooms, Facebook, blogs, online petitions, Wikis, podcasts, viral Flash films, cell phones, and text messaging, nonprofit organizations and activist movements have educated, agitated, and organized their constituencies and the broader public. Of course, nonprofit and activist organizations still inform and mobilize their membership via "snail mail" or print mailings, letter-writing campaigns, print newsletters, paper petition drives, phone trees, face-to-face organizing meetings, and door-to-door canvassing. Street events such as protests and marches are still *de rigeur*, albeit often digitally managed via meet-up software that allows people to log on and to see where vigils, marches, and actions are planned in their neighborhoods. As Howard Rheingold (2002) argues, mobile communication technologies have enabled "smart mobs" to form: Collections of "people who are able to act in concert," to share information, organize, and take action, "even if they don't know each other" (p. xii). To provide just one example of "smart mobs" in action, Heeman Kim and William Faux's chapter in this collection recognizes the crucial role that cyberactivism played in the 2002 presidential election in South Korea, during which "Internet users aggressively constructed their political and social agendas" (p. 158). The Web is literally rewiring and rewriting social movements around the world.

These digitally organized mass mobilizations have supplanted the centralized manifesto-led political organizing that characterized the 1960s and 1970s. As Naomi Klein (2002) points out in *Fences and Windows*, the communication technologies that shaped campaigns such as the "Battle for Seattle" (the World Trade Organization [WTO] protests attended by labor unions, Zapatistas, anarchists, and activists dressed as sea turtles) have shaped the movements "in [their] own Web-like image" (p. 17). Howard Rheingold describes the November 30, 1999 "Battle for Seattle" as a "smart mob" in which "autonomous but internetworked squads of demonstrators . . . used 'swarming' tactics, mobile phones, Web sites, laptops, and handheld computers" (2002, p. 158). "Thanks to the Net," writes Klein, "mobilizations occur with spare bureaucracy and minimal hierarchy; forced consensus and labour manifestos are fading into the background, replaced instead by a culture of constant, loosely structured and sometimes compulsive information swapping" (2002, p. 17). Whereas the activist network was once the snail-mail address list, it is now the Facebook cause list or activist social network like Think MTV. Such networks make possible what Clay Shirky (2008) calls the power to "organize without organizations" and the power to reconceive models of political participation (p. 171). As Shirky notes, "Political action has changed when a group of previously uncoordinated actors can create a public protest that the government can neither interdict in advance nor suppress without triggering public documentation" (p. 171).

Volumes, of course, have been written about the successes of cyberactivism, with example after example of success being cited—a virtual cyberutopia. Yet even as we extol the possibilities of cyberactivism and digital communication technologies, we must recall, as Tehnmozhi Soundararajan (2003), Executive Director of the Third World Majority (TWM), reminds us, "when you're online a whole lot of people are not in the room" (cited in Boyd, 2003, para. 31). Although use of the term *digital divide* can be reductive, it can have strategic value in describing the material conditions of access for many people. The TWM, dedicated to improving Web access for all the world's peoples, identifies the digital divide as one of empowerment: "The gap between people and communities who can make effective use of information technology and those who cannot" (2002, para. 1). The TWM Web site observes that these inequalities are not new. Rather, they are a "continuance of already existing divides of race, class, and gender disparities that are firmly institutionalized in the U.S. and other Western nations rather than a new and isolated phenomenon" (para. 1). As has been pointed out in government studies of digital divergence in the United States and analyzed in this collection's introduction, when a household and a neighborhood have no or insufficient wire service and when free Internet access is available only through a public library or public school and involves long waits and limited hours, whole groups of people are distanced from significant political, economic, and social opportunities. A lack of public participation in "social debates and political discussions" by specific groups means there is no "organized collective effort to challenge and reform these sectors" (para. 3). Organizations such as the TWM emphasize that the way to address the negative effects of economic disparities on digital advocacy is for those who have traditionally been excluded to shape and to reshape the very terms of use. TWM advises that "we have to look at technology critically and understand the social, political, and economic forces that went into creating the technology that we use (and mostly likely built) and then look at how the voices and visions of all of our peoples can be in the center of building long-term sustainable models for technology that contribute to the movement for global social justice" (para. 5).

A need to make explicit connections between social justice and technologies is felt by many teachers of professional communication. Further, pedagogical strategies that use technology for community action may surface divergence considerations unique to cyberactivism—topics that communication instructors may use as opportunities for encouraging critical literacy skills. Tony Scott (2006) has reviewed technology instruction methods that cast a professional writer "as a politically/ethically conscious social actor, not only reading and responding to the beliefs and expectations that define particular communities, but potentially changing them" (p. 231). This pedagogical approach, Scott explains, emphasizes how professional writers' social agency can be enhanced through the strategic use of technology (p. 231). Anthony Wilhelm's *Digital Nation* (2004), which asserts technology's potential for democratizing economic and social systems (p. 10), describes one student team's use of Web authoring software to create a Web site encouraging ecological awareness (p. 26). And Melinda Turnley (2007) explores how "service learning and client-based" technical communication curricula, which like critical literacy "focus attention on the values and assumptions embedded in our discourse and practices" (p. 107), can be part of an ethically oriented pedagogy. After completing a service-learning experience, her students were asked to write a reflective essay that highlighted, among other things, "the position of the client organization in the community and the significance of the social issues that it addresses" (p. 119).

The technology-justice relation model is, however, fraught with challenges insofar as the "informatics of domination" already are at work in digital environments (Haraway, 1991, p. 150). In *The Politics of Information*, Marc Bousquet (2003) refers to the Internet as a "managed and regulated space," a space of "commercial and state ideology, as well as a plethora of direct controls—an explosion of gateway structures, surveillance procedures, copyright protections, trademarks, standardization, advertising, certification processes, and other markers of intellectual processes"—in short, a place of "technocapitalism" and "cybernetic governmentality" (p. vi). Bousquet's cautionary tone is noteworthy, but he argues that we must move beyond the binary rhetorics of cyberutopianism or cyberdomination and begin to form "oppositional consciousness and alternative solidarities" (p. vii), the kinds of critical literacies and participatory practices that the TWM advocates and that many scholars and teachers of professional communication support. Bousquet asks, "How can students and citizens respond to this informatics of domination [Haraway's term] except with an answering informatic practice" (p. vi)? As an attempt to help students understand how they can make use of information technology, not as a transparent screen for the movement of information, goods, and services to consumers or as a place to go to find answers, but as a critical tool for knowledge construction, interpretation, and action in the world around them, the pedagogy described in this chapter offers one such "answering informatic practice."

CRITICAL LITERACY AND AGRICULTURAL LITERACY IN THE CYBERACTIVIST CONTEXT

It has become commonplace for professional communication teachers interested in digital media to argue that students must approach digital media as critical users. Doreen Starke-Meyerring, Ann Hill Duin, and Talene Palvetzian (2007), for instance, describe "critical reasoning from multiple perspectives" (p. 154) as one of the primary goals of the Global Classroom Partnership learning experience for digitally linked technical communication students in Russia, Sweden, and the United States. And Michelle Eble and Lynee Lewis Gaillet (2004) review in communication pedagogy literature various understandings of technological literacy, including their relations to critical literacy, to advocate for a "civic-minded and rhetorical approach to professional and technical communication courses and program design" (p. 350). While I sympathize with these and other professional communication teachers' emphasis on and conceptualizations of critique, I worry that we often do not identify the bar for "critical use" of digital media. Referencing the professional communications produced by educational organizations such as the NCTE, Cynthia Selfe (1999) warns that although these groups' "standards documents now . . . assume the necessity of computer use by communicators in the 21st century, they do not provide adequate guidance about how to get teachers and students *thinking critically about such use*" (p. 419, emphasis in original).

In the professional communication classroom, critical literacy work is a central concern as we work with students who are likely to be well versed in technology use but less so in the importance of critically examining the deployment of technology toward specific societal ends. Cyberactivism forms such as *The Meatrix*, with its appealing and strategic way of introducing students to the debates over factory farming, can provide case studies of the potentials and pitfalls of new media as advocacy resources and thus can constitute opportunities to enact the analytical skills comprising a rigorous critical literacy.

Literacy, as Deborah Brandt (1998) defines it, is a resource that is "economic, political, intellectual, spiritual—which like wealth or education, or trade skill or social connections, is pursued for the opportunities and protections that it potentially grants its seekers" (p. 5). Critical literacy, contends Ira Shor (1999), "challenges the status quo in an effort to discover alternative paths for self and social development" (para. 2). Critical literacy allows us to analyze and to rethink the practices of everyday life and encourages dissenting perspectives that allow people to connect "the political and the personal, the public and the private, the global and the local, the economic and the pedagogical for rethinking our lives and for promoting justice in place of inequity" (para. 2). Critical literacy, then, "can be thought of as a social practice in itself and as a tool for the study of other social practices" (para. 27). Following the lead of these scholars, I see critical literacy as a way to help students and communities move toward creating knowledgeable, literate citizens, prepared with the "knowledge and skills for social and environmental justice" (Andrezejewski & Alessio, 1999, para. 18).

The kinds of literacies needed in a globalizing world are not only the widely touted information or digital literacies, but also a critical literacy that allows citizens to understand specific networks of communities and global linkages. As Klein argues in *Fences and Windows,* understanding the concept of global linkages is "about recognizing that every piece of our high-gloss consumer culture comes from somewhere. It's about following the Webs of contracted factories, shell-game subsidiaries, and outsourced labor to find out where all the pieces are manufactured, under what conditions, which lobby groups wrote the rules of the game and which politicians were bought off along the way" (2002, p. 30).

While Klein's comments about understanding "global linkages" address alternative globalization movements specifically, her comments can be applied to questions of digital literacies and agricultural literacies as well. In a consumer-oriented, largely urban society, critical literacy about issues of food production is an especially important site of intervention, for the way that food is produced, marketed, and sold affects everyone on a daily basis and has a daily environmental impact on the planet. To illustrate how students can come to understand a complex issue such as food production and factory farming critically, I will contextualize *The*

Meatrix and then explore how this viral Flash activist film as a specific form of cyberactivist discourse might be considered in the communication classroom.

VIRAL FLASH ACTIVISM

In November 2003, links to *The Meatrix* jammed the e-mail inboxes of computer users across the globe. A snappy 3-minute, 47-second Flash film mimicking the plot and characters of the cult-classic film *The Matrix*, *The Meatrix* details the "hidden truth" about our food supply: That the relatively small family-farms of our national unconscious have been replaced by factory farms, or Concentrated Animal Feeding Operations (CAFOs) run by large agribusiness corporations. Produced by the new-media company Free Range Studios, for the Global Resource Action Center (GRACE, also known as the Sustainable Table), a nonprofit organization that resists factory farming and promotes sustainable agriculture, *The Meatrix* provides students with an alternative lens through which to survey the social, economic, environmental, and cultural effects of factory farming on rural communities and on us all, since we all are consumers of food. *The Meatrix* also provides an interesting example of new-media organizing and critical literacy in action, as it models viral Flash activism as a form of cyberactivism.

Viral Flash activism deploys short Macromedia Flash films that convey an organization's or group's message through a narrative meant to "inspire, amuse, and activate" a "target audience" (Free Range Studios, 2006, para. 3). Viral Flash activism films often make use of humor and pop culture references and thus are an effective cyberactivist organizing tool for many audiences, including many college students. At the end of a viral Flash activist film, viewers are presented with a "clickable call to action," which urges them to get involved by, for example, signing a petition, joining an organization, participating in a specific campaign, or donating money. The term *viral* refers to the fact that the Flash film is distributed quickly via a membership list and then mailed by members to friends, family, co-workers, and acquaintances. Viewers do not have to go a theater to see the film; instead, it is spread like a virus through the online community, mailed, and then forwarded and reforwarded to maximize its impact. Bloggers link to the film, as do the Web sites of nonprofits and political advocacy groups, thus accelerating the film's impact on the originating organization (Free Range Studios, 2006, para. 4).

The medium of viral Flash activism has caught on; and Flash advocacy films such as the Organic Fair Trade Association's 5-minute film *Grocery Store Wars*, a spoof of *Star Wars* and a critique of genetically modified foods, are becoming increasingly commonplace. *The Meatrix* also has sparked two spinoffs: *The Meatrix II*, which is the story of the industrial dairy system, and the *Meatrix II½*, which narrates the meatpacking industry and was linked to the in-theater release of the film *Fast Food Nation*.

Free Range Studios claims that viral Flash activism used as an organizing tool for nonprofit organizations has the potential to quadruple the response these organizations normally get from direct print mailings (para. 7). Free Range also claims that while direct mail campaigns usually bring about a 1.5% return, or 150 people for every 10,000 contacted, viral Flash films can very significantly increase that response rate. In short, according to Free Range, viral Flash activism is a "more cost-efficient, sexier, and faster" medium "than mail, radio, or television" (para. 4). Kalle Lasn, the founder of Adbusters, notes that although television is still the priority medium, Flash films serve as a "strategic pincer" in calling people's attention to an issue in a digital environment (qtd. in Shreve, 2001, para. 8). For example, when Free Range made a 90-second movie for the Brady Campaign to Prevent Gun Violence, the movie was viewed 35,000 times and resulted in 1,700 people signing up for the Campaign's newsletter. As a result of viewing the movie, 3,000 individuals sent a letter of protest to NRA then-President Charlton Heston (Free Range Studios, 2006, para. 8). The success of this mode of technological organizing is tied to a viewer's digitally mediated choice to get involved in an issue. The viewer decides to watch a Flash film, follows a link to the film, watches a short movie, and follows another link to access more information, thus "Webbing" into a series of further resources. It's easier than writing down a phone number viewed on television and easier than calling into a phone center. Digital users are able to get involved almost instantaneously, thus increasing the chances that viewers will take action (para. 18). "Going viral," of course, is not a new phenomenon or limited to the Internet. One of the key commentators on "media viruses" in the 1990s, Douglas Rushkoff (1994), described that widespread popular culture phenomenon:

> Media viruses spread through the datasphere the same way biological ones spread through the body or a community. But instead of traveling along an organic circulatory system, a media virus travels through the networks of the mediascape. The "protein shell" of a media virus might be an event, invention, technology, system of thought, musical riff, visual image, scientific theory, sex scandal, clothing style, or even a pop hero—as long as it can catch our attention. Any one of these media virus shells will search out the receptive nooks and crannies in popular culture and stick on anywhere it is noticed. Once attached, the virus injects its more hidden agendas into the datastream in the form of ideological code—not genes but a conceptual equivalent we now call "memes." Like real genetic materials, these memes infiltrate the way we do business, educate ourselves, interact with one another—even the way we perceive reality. (pp. 9–10)

In The *Viral Spiral* (2008) David Bollier argues that viral "refers to the way in which new ideas on the Internet can proliferate with astonishing speed" (p. 1). Bollier's analysis, nearly 15 years after Rushkoff's work, shows how viral technology has endured as a compelling digital form—seen for example in increased commercial use of "viral marketing" to ensure the success of a product or service (p. 131).

THE MEATRIX: A CASE STUDY IN THE LIMITS AND THE POTENTIALS OF VIRAL FLASH ACTIVISM

Starring Leo (à la Neo from *The Matrix*), a simple cartoonish pink pig, and the black-trench-coated cow Moopheus (à la Morpheus), *The Meatrix* opens with Leo's idyllic, slow-paced life on a family farm. An amiable farmer in overalls fills a pig trough with slop. Just as the blissfully unaware Leo is ready to tuck into his trough, a cow in sunglasses steps out of the shadows and says "Psst, Leo." After introducing himself as "Moopheus," he asks if Leo wants to know what the Meatrix is. Leo, of course, assents, and Moopheus explains in a deeply resonant voice (resembling that of actor Laurence Fishburne from *The Matrix)*, "The Meatrix is all around you, Leo. It is the story we tell ourselves about where meat and animal products come from. The family farm is a fantasy, Leo. Take the blue pill and stay here in the fantasy. Take the red pill, and I'll show you the truth."

Leo chooses the red pill, and the idyllic farmyard disappears around him, replaced by the close confinement feeding shed of a hog factory farm. "Welcome to the real world," intones Moopheus. "In the mid-20th century, greedy agriculture corporations began modifying sustainable family farming to maximize their profits at great cost to both humans and animals. Factory farming was born."

Leo watches with dismay as small family-farms are systematically flattened by tentacle-waving robots emblazoned with the logo "Agri-corps." In place of the red barns symbolizing family-farms, the agri-robots erect factory farms: Long steel sheds that confine animals to small stalls where they cannot move or even turn around. As the turn from family farming to factory farming for reasons of efficiency and profits is narrated, the hazards associated with it appear as bulleted talking points in the left corner of the screen: Animal cruelty, antibiotic-resistant bacteria, massive pollution, and destroyed communities. The film illustrates each problem, for instance, by stating the fact that 12 million pounds of excrement are deposited by animals in factory farms (GRACE/The Sustainable Table, 2006), and by presenting images of chickens being debeaked.

As Leo is led out of ignorance into awareness of the situation around him, he transforms from a clueless pig into a factory-farm fighter, executing stop-motion karate sequences that are borrowed from *The Matrix*. Finally, Leo dons the ubiquitous black sunglasses that indicate his passage over to the "resistance" movement against the factory farming "Meatrix." At the end of the film, Moopheus turns directly to face the computer user, stating, "We are going to spread the word. But it's you, the consumer, who has the real power! Don't support the factory farming machine! There is a world of alternatives." In his hoof, he holds out a red pill and says, "Click here and I'll show you what you can do to escape the Meatrix."

The film's "virus shell," or that aspect of the film that quickly engages the viewer's attention, is the parody of the cult-classic status of the film *The Matrix*. The punning title also makes the film intriguing to potential viewers, sparking in them a desire to understand how *The Meatrix* differs from *The Matrix*. And because links to the film are mailed to associates or friends, there is a higher likelihood that viewers will click, link, and forward, thus expanding the possibilities for the film to reach a wider audience. In this way, the Flash film provides viewers an opportunity to reach out to friends with a "cool" and humorous film without necessarily being fully aware that they are spreading a very political message for a nonprofit organization fighting factory farming.

In addition to being funny and clever, steeped in the "virus shell" of pop culture references to the *The Matrix*, *The Meatrix* is an effective organizing and educational tool for four reasons. First, it demonstrates a problem that many of our students and the general public may be unaware of—factory farming. *The Meatrix* narrates briefly, yet persuasively, the shift from family farming to factory farming and offers a simple and engaging animated narrative about why it happened: Greedy agriculture corporations or corporate agribusiness flattened family farms in the name of profit and efficiency. There is, of course, more to that story, which I will return to in a moment.

Second, *The Meatrix* creates a sense of identification in the Burkean sense. The film urges viewers to understand how we, at the grocery store, are a bit like the unsuspecting Leo at the slop trough. We might think and hope our food is being produced safely and sustainably, yet we really don't know; perhaps, on some level, we may not care or even want to know. We are in the fantasy world, or "Meatrix," of the family farm and blissfully unaware of how our meat and animal products are produced. After all, the commonplace cultural narrative of our food supply is that modern technology and efficient systems of production and distribution have liberated us from the need to grow or raise our own food and from the burden of being aware of how it reaches our tables.

Third, the film helps viewers come to terms with some of the costs of industrialized agriculture: Costs to animals who are frequently mistreated and abused, costs to the environment because CAFOs' overflowing manure lagoons are contaminating the air and water supply in rural communities, costs to rural communities because jobs are being compromised and family farms are being lost, and costs to consumers because antibiotic-resistant bacteria are being developed and our meat supply is no longer safe.

Finally and most importantly, the film issues a call to action urging consumers to fight back through supporting sustainable food systems. The clickable link to the GRACE/Sustainable Table Web site takes us to resources and information about how consumers can buy sustainable food from local family farmers. Although many aspects of the family farm-to-factory farm story are of course simplified in *The Meatrix* (for the medium of Flash film dictates a very concise message), the film does not allow us to wallow in our lack of awareness. Instead, it advocates that we become critically and "agriculturally literate" by recognizing our stake in how food is grown and produced. What is more, the film helps us understand the social, ethical, and environmental costs of the agro-industrial complex. The "clickable call to action" entreats us to become engaged, to learn more, and to seek food alternatives.

The call appears to be effective. GRACE reports that *The Meatrix* "has been translated into more than 30 languages and is now the most successful online advocacy film ever, with well over 15 million viewers worldwide" (2006, para. 1). But my interest in *The Meatrix* does not center on its utopian success as an example of cyberactivism. What interests me about the Flash film is how it offers viewers a way of seeing—and of *not* seeing—their own agency in the processes of food production and consumption and, subsequently, what the film can do to inspire and motivate critical literacy about food production among students.

Clearly, *The Meatrix* works to sponsor a critical literacy moment, moving us beyond mere agricultural literacy (simply knowing how food is grown) to what Jacqueline Edmondson (2003) calls a new *alternative agrarian literacy*, or an understanding of the environmental, social, economic, and political consequences of the food system (p. 15). At the same time, the film glosses over the complex histories and institutional and ideological narratives of agriculture—narratives of progress through science and increased efficiency—that have brought us to our current relation with the food system. When *The Meatrix* is used as a case study in the communication classroom, the process of exploring and inquiring into such cyber-activism may help students understand the possibilities and limitations of texts such as Flash films and nonprofit Web sites.

The Meatrix proffers an image of a food consumer who has been duped by the system rather than of a consumer who has rationally and willingly bought into the food system's narratives of efficiency and cheapness. In fact, the film lets viewers off the hook with its overly simplistic narrative of the rise of factory farms. Herein lies the difficulty of using Flash activist films to promote social change and critical agricultural literacy. By their nature, Flash activist films are short narratives and cannot tell the "whole story." Salient in *The Meatrix* is the story of greedy agribusiness corporations flattening family farms. What is not so clear is how farmers and consumers participated in this project of "industrializing" the family farm and how they were aided and abetted by agricultural corporations, new technologies, government agencies, and, ironically, land-grant institutions, the universities where many of us are employed.

The narrative line in *The Meatrix* therefore strips the viewer of agency to some degree and simplifies the systems of production, distribution, and cultural narration that have enabled the factory-farm system. While we may identify with the factory-farm fighters Leo and Moopheus, we are discouraged from identifying with the Agri-corps robots flattening farms and stamping the "Happy Farms" label on our food. The meme (Rushkoff, 1994) to pay attention to in *The Meatrix* is not the Meatrix itself, as implied by the film's narration, but the machine that brought the Meatrix—the Agri-corps-emblazoned robot. All fans of science fiction know that the only way to destroy the renegade robot is to destroy its knowledge core—its "brain." If the hero or heroine can destroy the brain, he or she can destroy the robot and stop its destruction. The problem with the tentacle-waving robot of *The Meatrix* is that its core is made up of knowledge workers whom we do not see in the film. If we opened up the Agri-corps robot, we would find the "loathsome" agribusiness executives, the "men in black" corporate suits portrayed in *The Meatrix II* and *The Meatrix II½*, but we would also find family farmers and a host of knowledge workers: Agricultural economists, professors at land-grant institutions (including teachers of professional and technical communication), geneticists, policymakers, lawyers, nutritionists, and others. We have met the enemy, and, in some ways, it is us, the knowledge workers of this society deploying the twin resources of science and technology in the names of capitalism and of efficiency, productivity, and profitability.

Thus, the Flash film insufficiently implicates us and insufficiently highlights our willingness as consumers to "take the blue pill" and to continue dwelling in the

Meatrix for the sake of convenience, technology, and economic efficiency. The film does not show how the broader U.S. public has been as mesmerized as Leo with images of plenty, 24-hour access, and year-round availability of all foods at a cheap price, and it does not acquaint us with the challenges we might have seeking alternatives to the meat monopoly. The reason for this absence is ironic: The elements that make factory farming so central to food-production systems—knowledge work and technology—are the same elements that enable viral Flash activist films like *The Meatrix* to exist. Narratives of science, technology, progress, efficiency, and endless plenty are all at the core of the Agri-corps robot. The film also makes invisible the complex of farmers and workers that tend to CAFOs and that staff meatpacking plants. *The Meatrix II½* continues the story and does address the meatpacking industry and workers, but the story of the labor of factory farms and the knowledge workers who created them are largely lost in *The Meatrix*.

While there is no doubt that *The Meatrix* helps viewers draw some preliminary conclusions about the basic tenets of the factory-farm system, the overall message about factory farming is too simple—a fact acknowledged by the media designers of the film, for Flash films are usually embedded in and attached to other Web resources that offer depth and breadth on social and political issues. For the film to achieve its desired effect of directly affecting consumer behavior, the viewer certainly will have to interact further with other digital media—the GRACE/Sustainable Table Web site, for instance—in order to learn more about the history, causes, and effects of factory farming. Yet the Web resources that the film links to need to be embedded more broadly in a larger set of texts and contexts that explore more fully the implications of the issue of factory farming and that do not necessarily depend on digital access. Through exploring the film and the Web site's resources and links, professional communication teachers can help students to appreciate digital media for the resources they provide, but also to network and to web across sources, to dig much deeper, to ask more challenging questions than simply "How is our meat grown?" and to explore a wider variety of sources, including traditional print or multimedia documents, thereby situating the film in a larger complex of virtual and actual worlds.

USING *THE MEATRIX* TO ENGAGE CYBERACTIVISM CRITICALLY IN THE PROFESSIONAL COMMUNICATION CLASSROOM

A helpful way to begin exploring the rhetoric of *The Meatrix* in class is to ask students to follow the links provided by the GRACE/Sustainable Table host Web site and to consider the ethos of these related organizations. Students should also be asked to network beyond the social worlds strategically constructed in the Web sites that a Flash film offers. To help students think beyond this carefully packaged mix of sources, I show the film in conjunction with a discussion of Eric Schlosser's (2001) book *Fast Food Nation*, assigning as a companion piece to the film the fifth chapter of his book, which addresses the transformation of the family farm. The chapter explores the difficulties faced by Hank, a cattle rancher in Colorado who is being squeezed by the suburbanization of ranch lands, corporate concentration, and myriad social and economic forces. The chapter closes with his suicide, a shocking revelation in the book, but an important commentary on the rash of farmer suicides in the wake of plummeting crop prices in the 1980s and 1990s.

After discussing *The Meatrix* and *Fast Food Nation*, I show students specific Web sites where they can view different images of CAFOs: www.factoryfarm.org and the Sierra Club Web site's links on factory farming. I ask them to consider how the images of factory farming are visually presented and to what effect. How do those images compare with those provided in *The Meatrix*? Subsequent class discussions are devoted to understanding how and why factory farming developed. I ask students to conduct research on local library databases and specific Web sites about what factory farming is, when it started, and why it developed into a widespread system of production. I also ask that they research and read sources from a variety of perspectives and explore the rationales for factory farming.

My goal in directing students to a range of sources beyond those provided by the film and the GRACE/Sustainable Table Web site is to ask them to consider how the film and its network of sources gives sufficient context—or not—for the public to become informed about factory farming. I ask students to consider the definitions, concepts, and historical narratives around factory farming and the ways in which they are constructed through a network of sources and research communities that pose specific arguments for caring about the issue—moral, ethical, economic, and environmental. I also ask students to consider the roles that public policy and industry play. I encourage them to consider carefully the Sierra Club and the Citizens' Environmental Coalition *Report on Factory Farms in the State of New York* (see Schade, 2005), which documents the rise of factory farms in upstate New York, the backyard of Syracuse University, where our class takes place. By asking ourselves who can access organic and sustainably produced foods, and who cannot and why, we also consider the feasibility of organic and sustainable farming as an alternative to family farming.

Finally, I ask who can view *The Meatrix* and who cannot—how wide and deep its reach is as a form of cyberactivism. Given the complexities of digital divergence, who is in and who is out? Not surprisingly, my students note that viral Flash activism is most likely to reach middle-class people in economically developed nations with modems and high speed Internet access, although cell-phone technology is changing this situation to some degree. Given the access challenge that exists for so many people, I ask students to consider how other media or "on the ground" organizing projects can convey messages about factory farming that will reach a wider audience. One such pathway we investigate is urban public-school projects such as "farm-to-cafeteria" and edible schoolyard programs, through which students access locally farmed foods and grow some of their own food as part of the science curriculum.

As one might imagine, class discussions of factory farming can be volatile. *The Meatrix*, after all, is a dystopian narrative, not the happy story of Old MacDonald's farm or The Farmer in the Dell. Some students are strongly affected by the story narrated by *The Meatrix* and the graphic images they see on Web sites of beef and dairy cattle, veal, chicken, and swine living in CAFOs. Those most affected by the images usually appeal to the ethics of animal rights and sustainable farming, free-range meat and chicken, cage-free eggs, and organic milk. Other students are concerned with issues of ample food supply and cost-savings to consumers. They argue that cattle, pigs, and chickens are domesticated animals born for slaughter and that our own populous species can be fed only by means of a food system characterized by efficient conditions and affordable prices. Other students point to public policy and to how CAFOs pollute the environment, costing citizens millions of dollars every year. They cite information that indicates that CAFOs do not necessarily produce safe food, citing potential food-supply problems associated with avian influenza (bird flu), *Escherichia coli*, and bovine spongiform encephalopathy (mad cow disease). Discussions of oligopoly in the meat industry also arise. These arguments lead to interesting discussions of the regulation of CAFOs, which environmental activists have labeled "farms with a permit to pollute." We talk about the role of the Department of Natural Resources, the Food and Drug Administration, the Department of Agriculture, and the food safety and inspection systems.

Even though my students do not tend to agree on the argumentative stance that they and other class members should take toward factory farming, they do agree that public information and school education campaigns are necessary to address the problems and questions posed by the practice. Thus, it is evident that *The Meatrix* creates a sense of mutual identification where we do not just sympathize uncomfortably (those poor animals—*mmmm*, those burgers), but we see the larger dimensions of the problem and their far-reaching effects on the food system, the environment, and our communities. We are implicated, to some degree, in the problem of factory farming and are invited to take action. Whether we are rural residents or residents of suburban and urban communities, *The Meatrix* makes it clear that the issue of factory farming touches us all, and an increasing awareness and movement toward critical literacy is the most appropriate action to take. Yet to take advantage of the message embedded in *The Meatrix*, the agency of the viewer must be acknowledged fully, both in his or her own choice to remain embedded in the factory-farm system or to break out and explore different options even as issues surrounding digital divergence must be explored as part of that context: If the message does not reach the audience, then exclusion has been enacted. Unpacking advocacy texts while honoring an audience's agency highlights the complex balance that professional communication students must maintain when creating their own documents, which will ideally reflect "the concerns of all stakeholders, not just the powerful, who have an interest in their product" (Cook, 2002, p. 16).

CONCLUSION

As cyberactivist discourses continue to drive much of the advocacy rhetorics of social change in the coming decades, teachers of professional communication can investigate with our students the resources and rhetorical strategies provided by such discourses, as a method for fostering critical literacy. At the same time, we would do well to slow down and engage in our own critical literacy work, forging our own critical informatic practices (Bousquet, 2003) by investigating what digital media provide and what they do not, and considering who they do and do not reach. As Laura Gurak and Ann Hill Duin (2004) argue, we must avoid the "false binary" with respect to technology: We must avoid viewing digital environments as boon or bane, "evangelizing about the power of the Internet and digital technologies on the one hand or criticizing and dismissing them on the other" (pp. 187–188). The best antidote to such binary thinking is an answering informatics of practice that encourages critical literacy work.

As Moopheus says in *The Meatrix,* "Take the red pill!"

REFERENCES

Andrezejewski, J., & Alessio, J. (1999). Education for global citizenship and social responsibility. *Progressive Perspectives, 1*(2), 1–23. Retrieved May 20, 2008, from http://www.uvm.edu/~dewey/monographs/glomono.html#Education%20for%20Global%20Citizenship%20and%20 Social

Bollier, D. (2008). *The viral spiral: How the commoners built a digital republic of their own.* New York: The New Press.

Bousquet, M. (2003). The politics of information: An introduction. In M. Bousquet & K. Wills (Eds.), *The politics of information: The electronic mediation of social change* (pp. v–vii). Boulder, CO: Alt X Press.

Boyd, A. (2003, August 4). The Web rewires the movement. *The Nation*. Retrieved March 20, 2008, from http://www.thenation.com/doc/20030804/boyd

Brandt, D. (1998). Sponsors of literacy. *College Composition and Communication, 49*(2), 165–185.

Brooke, R. (Ed.). (2003). *Rural voices: Place-conscious education and the teaching of writing.* New York: Teachers College Press; Berkeley, CA: National Writing Project.

Cook, K. C. (2002). Layered literacies: A theoretical frame for technical communication pedagogy. *Technical Communication Quarterly, 11*(1), 5–29.

Eble, M. F., & Gaillet, L. L. (2004). Educating "community intellectuals": Rhetoric, moral philosophy, and civic engagement. *Technical Communication Quarterly, 13*(3), 341–354.

Edmondson, J. (2003). *Prairie town: Redefining rural life in an age of globalization*. Boulder, CO: Rowman & Littlefield.

Free Range Studios. (2003). *The Meatrix*. Retrieved March 4, 2010, from http://www.themeatrix.com/

Free Range Studios. (2006). *Forget about direct mail* Wayback Machine Internet Archive. Retrieved May 21, 2008, from http://web.archive.org/web/20060304230152/www.freerangegraphics.com/html/gallery/Flash.html

GRACE/The Sustainable Table. (2006). *The Meatrix: About.* Retrieved May, 20, 2008, from http://www.themeatrix.com/about/

Gurak, L., & Duin, A. H. (2004). The impact of the Internet and digital technologies on teaching and research in technical communication. *Technical Communication Quarterly, 13*(2), 187–198.

Hackos, J. T. (1994). *Managing your documentation projects*. New York: Wiley.

Haraway, D. (1991). *Simians, cyborgs and women. The reinvention of nature.* New York: Routledge.

http://dx.doi.org/10.2190/CWDC2

CHAPTER 2

Retracing the Footprints from Print to Digital: An Assessment of Textual Structure

Adrienne P. Lamberti

OVERVIEW

The response to a divergence from the norm may be reluctance and even resistance, and both surface in scholarly reactions to digital technology's effects on traditional argumentative structure in texts. Such reactions concerning the structural meeting place between print and digital arose during the author's recent efforts to write a "Webtext" for the journal *Kairos*. This chapter describes the effort to construct the text, then reviews scholarly responses to the prospect of textual restructuring as encouraged by digital forms. Examining assorted writing interests as they advocate their respective structural authorities offers a picture of the issues at stake in the larger phenomenon of discourse norming. Such responses include comprehension and composition processes invoked by digital media. The chapter also includes a disciplinary-specific analysis of digital technology as it has affected the concept of sole authority, exemplifying one manner in which technology-mediated structure and its impact upon authorship may transform rhetoric.

Debate concerning the ideal methods of arranging a text implicates anyone who happens to write, but is especially pertinent to writers/rhetors in scholarly and pedagogical contexts. That several arrangements will be promoted as best in any given rhetorical situation suggests the extent of the contestation regarding *dispositio*, or the "arrangement of that which has parts" (Aristotle, trans. 1970). Historically, the traditional argument structure of Western print texts has been characterized as exhibiting a purportedly linear line of reasoning, inductive or deductive, in support of a ruling claim. But debate over *dispositio* is being amplified as the ability of digital technologies to generate divergent forms of heretofore print texts becomes increasingly evident.

One implication of digital technology is its potential to reconfigure traditional strategic moves and structural conventions found in Western communication models. Digital versions of such texts may involve structures ranging from neotraditional (e.g., the PDF version of a peer-reviewed journal article initially published in print form) to divergent dynamic arrangements (e.g., a Web page with hyperlinks). Indeed, the ground has been readied for revisiting communicative norms—and for studying their transformation. John Schilb's *Rhetorical Refusals* (2008), for instance, suggests how writers' repudiation of expected rhetorical moves can both surface the warrants behind argumentation conventions and offer examples of alternative strategies for communication. Writers, scholars, and teachers who engage with the sophistry afforded by the digital may find options for communication that are intellectually freer and more open, even playful.

Still, responses to a divergence from the perceived norm are often reluctance, if not more pointed acts of resistance, and digital technology's effects on traditional argument structure have given rise to both types of response. Efforts to generate consensus regarding preferred textual structure engulf all groups associated with writing and rhetoric. Representatives from various scholarly and pedagogical interests have demanded that what counts as acceptable argument structure be completely unrestricted, heavily policed, and everything in between. This contested territory offers anyone who brainstorms, writes, designs, or just thinks an opportunity to better understand the impact of digital technology on the establishment of "correctness" generally. Patrick Dias, Aviva Freedman, Peter Medway, and Anthony Pare (1999), citing Fairclough (1995), provocatively describe this pressure as the "technologization of discourse, a process of intervention in the sphere of discourse practices with the objective of constructing a new hegemony in the order of discourse of the institution or organization concerned" (p. 123). Examining various responses as they make claims to authority can help surface the players involved in negotiating the standardization of discourse, an investigation of special significance whenever one consensus regarding correct structure seems to have achieved a precarious dominance.

This chapter explores one instance of the divergences afforded by digital technology and their transformation of rhetoric, by offering an evaluative snapshot of the heavily contested territory of traditional argumentative structure; in short, the chapter acts as one scholarly and interdisciplinary approach to a central topic, investigating how such divergences might become heuristic opportunities for writers/ rhetors. Beginning with a description of my efforts to construct a text for the Web, I review scholarly responses to the prospect of textual restructuring as brought on by digital forms, including its impact on comprehension and composition processes. Further, responses to these divergences as offered by the field of information science suggest a critical lens through which we might understand digitally informed authorship and architecture.

RESISTANCE IN REACTION: CHALLENGES TO READERSHIP AND RHETORIC

Resistance to divergence surfaces across industry as well as academe. For example, the annual report of the corporation Wire and Plastics Products (WPP), in anticipation of its 2005–2008 quadrennial fiscal cycle, repeatedly makes reference to "fragmentation," or the increasing unpredictability of users' acts of production and consumption as an outcome of new media proliferation. "Our practice," states the report, "will continue to be challenged by the volatility of the media communications landscape. One can scarcely pick up a trade or general market journal without seeing an article that speculates on the impact of fragmentation, TiVo/DVR, VOD, mobile platforms and other changes on . . . communications effectiveness" (Gotlieb, 2004, p. 44). The report reassures readers that WPP is working to better understand and to anticipate the new, digitally mediated, ad hoc communications being generated and demanded by consumers.

Commentary on the idea of a message's being scattered across digitized textual shards, including various forms of new media, may be provoked by the at-times discomfiting shift in experience that occurs when readers encounter a fragmented argumentative arrangement. Digital technology complicates, in ways that scholars are grappling to name, traditional assumptions about the processes by which communication shapes audience perceptions. James P. Zappen (2005) suggests that "studies [of the digital] are challenging the view that associates rhetoric exclusively with persuasion, a view that has persisted for more than two millennia" (p. 321). His review of literature regarding digital rhetoric includes, among other considerations, how "we have barely begun to describe and catalog" emerging ways the digital negotiates Web authors' and users' respective choices during their encounters with technology—a phenomenon that has been termed "rhetorical convergence" (Fagerjord, cited in Zappen, p. 322). Barbara Warnick (2005) examines how "screen-based interfaces" have significantly enabled users to individually customize electronic content (e.g., via defining Web browser settings; p. 329); each user therefore can craft a unique online experience (p. 330). Ultimately, Warnick argues that a user's experience with the digital is so qualitatively different from that with print that critiques of communicative effectiveness must break away from being "printcentric" (p. 327). In her view, what historically was criticism examining "one-to-many communication in single-authored writings" cannot be transferred appropriately to the digital, a medium characterized by user participation and multivocal discourse spaces (p. 332).

In such instances, the digital may act as a "disruptive technology" (Christensen, 1997). Not surprisingly, some digital technologies are being normed actively as a result of writers' and readers' discomfort with ambiguity and the unfamiliar. Brenton Faber's (2007) study of an e-mail conversation among a professional group's members chronicles how the technology is used to suggest, debate, and, especially, resist change to the manner in which e-mail posts are crafted. The members' interactions demonstrate how easily comparatively recent phenomena such as digital technology can be recruited to perpetuate historical understandings of what communication should and should not be. As Faber states, "Agents must have ways to limit competing discourses and ensure that forms/events of change are dissipated before gaining sufficient momentum to overtake other networks" (p. 239). If such discourse norming, a practice carried out presumably by all peoples at all times, itself is a type of "technologization" (Dias et al., 1999, p. 123), resistance to technology in the form of the digital suggests a complex situation indeed.

As one instance of critique in response to technology, this chapter examines the nature of rhetoric and its transformation through the digital by reviewing issues surrounding structural and strategic conventions. Technological or otherwise, the profound impacts of norming forces suggest consequences for larger argument strategy. By revealing structural and strategic conventions, those reconceptualizations prompted by the digital can help us compose and understand text in energizing ways, as illustrated in the following scenario.

A WEBTEXT > ABOUT A PRINT BOOK > WITH A DIGITAL DESIGN

On assigning Bruce Ross-Larson's *Writing for the Information Age* (2002) for an upper-level course on workplace report writing, it became apparent to me how the digital offers an opportunity to revisit conventional understandings of argument structure. Publisher W. W. Norton's blurb states that "whether you are writing a report, drafting email, creating a PowerPoint presentation, or building a Web site, this book shows how to use language that is easily accessible, never oppressive" (Writing for the Information Age, n.d., para. 1). My pedagogical intention for *Writing* was to explore how technology-mediated texts both support and challenge un"oppressive" efficiency models of business communication.

The text styles itself as promoting the spare, prescriptive writing philosophy set forth by Strunk and White's *Elements of Style,* yet visually arranges its information in a manner approximating a Web site, complete with ruling and subordinate layers of informational chunks mapped to one another throughout. In the bottom right corner of every recto is a design resembling the type of navigational menu that might be found on a Web site. For instance, the information in Chapter 3 on how to compose "engaging titles and subtitles" (2002, p. 3.1) is cross-referenced to Chapter 1 content advising the reader to "lighten what you write" (p. 1.1), "layer what you write" (p. 1.2), and "link what you write" (1.3), as well as to Chapter 4 content advising the reader to "know your topic, audience, and purpose" (p. 4.1) and to "start with your main and supporting messages" (p. 4.2). In keeping with the Web site metaphor, *Writing* is prefaced with a print version of a "How to Use This Book" sitemap, which distinguishes the layers and explains their interconnectivity (n.p.).

The students in my course were accustomed to a wired classroom, as the course met regularly in a computer lab, and were very experienced in using digital protocols such as Web interfaces and e-mail, including the more ad hoc, ephemeral networking behaviors (e.g., friending someone on Facebook) and fragmentary texts (e.g., text messages) enabled by such protocols. I thus assumed that the digital look of the Ross-Larson print text would prove an enticing novelty to students otherwise typically assigned traditional textbooks in their college courses. Instead, the majority of the class reported a distaste for the nonlinearity of the book's content. Believing that the need to physically move back and forth through pages of print to locate each reference (especially when compared with the little physical investment required during digital navigation, such as clicking a hyperlink) accounted for the student resistance, I investigated the complaint further.

In its arrangement, *Writing* violates linear exposition, a rhetorical strategy common to course textbooks. And I discovered that it was the *conceptual,* not visual, broken linearity that students found most disturbing. Trained to associate textbook genres with certain physical and organizational traits, the students viewed the book's mapped layers as representing a sort of informational scatterbrainedness that rendered the book's ethos suspect. The students' reaction not only prompted me to consider (in addition to questions concerning the book's efficacy for this specific course population) the blurry intersections between hard- and soft-copy texts, but also highlighted the evident one-way migration of certain conceptual constructs from print to digital form. That is, the visual familiarity of, for example, a printed text that has been formatted in Microsoft Word, scanned into a portable document format (PDF), and posted on a Web site may reassure readers of its reliability (despite or perhaps because of the likely vanilla design of such a document). Yet, as suggested by my students' reactions to the Ross-Larson text, there is an evident challenge in achieving the opposite: Convincingly transferring a digitally reminiscent conceptual construct to a print document. When print and digital texts are compared in such situations, conflicts between design and content may be discernable.

Intrigued by the experience, I began to consider the Ross-Larson text an opportune subject for a book review, and the nature of my topic an heuristic for locating a publication venue that similarly would be intrigued. Ultimately, I decided to attempt composing a digital text for the online rhetoric and technology journal *Kairos*. Exploring such conflicts is one aspect of the scholarly journal *Kairos'* mission: The editorial staff "invites each author or collaborative writing team to think carefully about what unique opportunities the Web offers" (*Kairos,* 2005, para. 1). Generally, *Kairos* identifies itself as a journal dedicated to "exploring the intersections of rhetoric, technology, and pedagogy," elaborating,

> In *Kairos*, we publish "webtexts," which are texts authored specifically for publication on the World Wide Web. These webtexts include scholarly examinations of large-scale issues. . . . With *Kairos*, we seek to push boundaries in academic publishing at the same time we strive to bridge the gap between print and digital publishing cultures. We further seek to bring forward and support the voices of those too often marginalized in the academy. (para. 1–3)

Here, three points are of note. First, the description suggests divergence in the spaces that result when rhetorical acts, such as traditional forms of research and writing, and rhetorical acts mediated by digital technology, such as hypertext production, interact. Second, the description implies that divergent texts may be generated from those spaces, and third, the description offers an opportunity as presented by these spaces and texts to front voices that somehow are silenced by more traditional textual forms.

A preoccupation with *dispositio* is implicated in all three points, making the Ross-Larson book, with its structural and navigational divergences, a relevant text to review. With these thoughts in mind, I envisioned a writing project whereby I would design a *Kairos* book review Webtext using an architecture that mimicked *Writing*'s structure. Planning the design of such a Webtext, however, suggested larger topics of contemplation beyond the book's specific content and *Kairos's* specific rhetorical situation:

- Should the Webtext discussion of the book's section 3.1 include a link, as it exists in the book, to relevant discussions regarding the first chapter?
- Or should the Webtext discussion of the first chapter's sections include links to discussions regarding section 3.1, in some sort of effort at chronology?
- Or both?
- Would designing the Webtext to exactly parallel the book's link structure prove confusing, in that several of the links would be neither chronological nor bidirectional (e.g., content discussing Chapter 3 would link to content discussing Chapter 1, but not vice versa)?
- Or would imposing a more chronological arrangement (e.g., with Chapter 1 content somehow acting as the "center" of the Webtext, as in a homepage or first sublevel Web page) violate the flattened hierarchy often present in and even expected of digital forms?

In working through these issues of structure, I found that my early research and drafting processes resulted in disconnected chunks of text, a not uncommon result of initial composing efforts in my experience, and not inappropriate content for the fragmented structure more often found in digital forms (see also Delagrange, 2009). Yet I was writing for an audience of fellow scholars in composition and communication, so I could not help but believe that my ultimate purpose while composing was to produce a cohesive discussion and analysis of *Writing* for readers who were used to and who expected a certain type of logocentric discourse. In other words, I was experiencing synchronous processes that at some moments were in competition: (a) spatial brainstorming efforts, which generate topics that often are not linearly arranged and therefore may be best represented by the flattened or even fragmented design architecture afforded by hypertext; and (b) efforts to construct a perceived linear line of reasoning, which typically is honored by argument structures traditional to print texts.

At the heart of my dilemma, then, was the question of whether the exigencies of such an endeavor would stymie inquiry, or whether its print-to-digital contradictions had the potential to energize composition. For instance, did the flattened, sequential, front-to-back, top-to-bottom, left-to-right physicality of the conventional Western print document in some way fail to satisfactorily represent more ad hoc intellectual efforts such as invention, whereas a hypertext writer's process decisions and more anarchic design more closely expressed and negotiated with scholarly reading/writing acts?

It seemed evident to me that structures enabled by the digital, where fragmented mappings and links and navigational structures are commonplace, offer exciting methods for exploiting the fragmented, nonlinear aspects of research, reading, writing, and revision—all respected historical acts endemic to scholarship. I was brought to recognize the many contradictions inherent in using a digital form (a *Kairos* Webtext) to assess for an academic audience accustomed to logocentric discourse (readers within the *Kairos* audience) the effectiveness of a print text (*Writing)* that visually modeled its nonlinear organizational structure upon a digital form (a Web site). A composing goal for my review therefore presented itself: How might I craft a Webtext to simultaneously honor the wider variety of organizational structures more enabled by digital forms (and by extension fulfill the target journal's mission) *and* accommodate the need to present a traceable thought process for readers of the journal, many of whom may have traditional expectations about scholarly argumentation? Would it be possible to compose an acceptable text that let these tensions stand?

The challenge in materializing the interactions between argument and arrangement has been addressed by others' discussions of digital technology. In an effort to understand how the movement from print to digital technology has influenced rhetoricians' attitudes toward *dispositio,* the remainder of this chapter reviews key responses to the digital restructuring of argumentative texts. Further, in helping to characterize the continuum of energizing textual options made available by the digital, I highlight one disciplinary-specific response to the impact of digital restructuring.

PRINT-INFORMED REACTIONS TO DIGITIZED DISPOSITIO: MOMENTS IN HISTORY

Historical precedent undergirding traditional argument structure restricts a writer's use of comparatively unconventional structures in print forms. But when content is generated by means of digital technology, a writer may have great latitude with arrangement. This section surveys selected arguments regarding arrangement across the disciplines, including familiar efforts to norm argumentative structure in print texts. Reactions to early, widespread play with digitized textual forms, which was facilitated by software programs such as HyperCard, illustrate how the digital disrupted conventional notions of argumentation in the early days of adoption and continues to generate opportunities for divergent writing pedagogies and scholarship.

Many early disciplinary reactions to structural disruption in digital communication were characterized by an effort to retrench and to achieve an overriding navigational structure. This effort was motivated largely by the expectation of audiences familiar with print; namely, that they would encounter information arranged in a manner so conventional as to seem transparently intuitive. Or, as Paul Delany and George Landow (1991) put it, "As long as the text was married to a physical media [sic], readers and writers took for granted three crucial attributes: That the text was *linear, bounded,* and *fixed"* (p. 3, emphasis in original). Mary Ann Eiler's (1998) comment that "hypertext was not imagined in its beginning as a way to support postmodernist theory" (para. 1) also suggests the tenor of early responses to the prospect of disrupted textual structure.

In any redefinition is the risk of anomie, so it is unsurprising that scholarly reactions to technological restructuring often reconciled paradigms of traditional argument structure with the potential offered by digital forms. Many studies were devoted to examination of the technology and to effecting a measure of certainty regarding digitally mediated structure. Computer scientists Richard Furuta and David Stotts (1990), for instance, describe how their initial focus on one specific hypertext system eventually led them to extrapolate its organization as a model for several if not most hypertexts. Furuta and Stotts lay out not just textually but graphically the scaffolding of their Trellis hypertext model (pp. 180, 182, 203). Notable about the authors' discussion is their frequent return to an emphasis on predictable structure both within and across hypertext systems, regardless of the level of conceptual and design specificity.

Similarly, during the digital's early appearances, scholars who focused on text composition also quizzed and corralled the implications of technology. Stuart Moulthrop (1991) offers reassurance for including in hypertext the acknowledgment of novel content (e.g., the *topoi* involved during a writer's use of the CARS model [Swales, 1990]), asserting that this inclusion in fact isn't so distant from traditional methods of connecting ideas: "Hypertext is nothing more than an extension of what literature has always been . . .—a temporally extended network of relations which successive generations of readers and writers perpetually make and unmake" (para. 19). Jay David Bolter's (1991) *Writing Space* suggests the use of digital graphics as rhetorical road signs, guiding users as they maneuver a site, as a means of synthesis between print and softcopy navigation devices (p. 61). Stephen Bernhardt's prediction in 1993 of nine ways in which text would be altered in its migration from print to digital argues that digitized text is not notably different from customary print designs (pp. 157, 162, 165). Essentially, Bernhardt's forecast of a text's technological incarnations reassure readers that the digital "borrows heavily on the evolved strategies readers possess for interacting with printed texts" (p. 166).

One case in point of the disciplinary reaction to the challenge posed by digitally redefined structure is provided by the popular 1990s program HyperCard. A software program debuted by Apple Computers in 1987 (and definitively discontinued in 2004), HyperCard was composed of a virtual "stack of cards—an endlessly spinning Rolodex"—allowing the user to perform data entry on each card (Belling, 2007, para. 2). Significantly, the content across a card stack could differ widely, as each card could exist as a component independent from the rest or be connected to the stack at the discretion of the user (Computer Chronicles, 1987). Discussions of HyperCard applications in the writing classroom consequently embodied a balancing act between embracing the exploded arrangement offered by a HyperCard stack and reiterating traditional structure. Marcus Otlowski (1998), focusing on the specifics of HyperCard–facilitated arrangement, encourages the use of computer software to make a text's structure visually apparent. Otlowski explains that many students who are nonnative English speakers are taught English in terms of sentence-level correctness; such an approach deprives them of exposure to the "structures that various text genres demand" (p. 420). Whereas Otlowski cites HyperCard as a pedagogical tool to help "the writer to create texts that can both stand alone and be integral parts of the main text" (p. 423), his support for the program concludes not by confirming the value of disconnected structure, but by encouraging teachers to use the program to encourage student writers to gather all their "stand alone" texts into a "complete document" (p. 423).

Granted, early scholarly response to the digital was not unanimous; moments can be spotted in the literature that support technology's potential to resist structural convention. John Slatin's (1990) "Order and Coherence in a New Medium" advocates that the advent of hypertext as a mechanism for knowledge making (in this case, during the composition of a text) is best represented by a rhetoric that "permits movement up and down the ladder of abstraction" (p. 870) to more appropriately represent what the text is attempting to do at each given moment. Slatin's claim is based on his assessment of how a reader's anticipation of "sequential" content arrangement and his/her consequent ability to make "predictions" based on a text's sequence—those traits endemic to the reading and writing processes of a fixed print text—are undermined by the comparative ease and frequency with which a hypertext's content can change (pp. 871–872). Johndan Johnson-Eilola's (1997) invocation of Derrida (1977) in *Nostalgic Angels* offers a postmodern approach to understanding digitally structured content. In his analysis of the interplay between structure and technology, Johnson-Eilola advocates that hypertext offers designers and users an opportunity for redefining the structure of content rather than merely transferring and imposing preexisting structures on emerging digital forms. In other words, the digital allows hypertextual structure to be understood as more than "absolute[ly]" centered or utterly decentered (pp. 158–159). This analytical opportunity has not only logistical implications (e.g., developing innovative hierarchies for navigating a hypertext) but ideological ones, in that traditional structure often compels a user to make certain decisions through top-down, one-way control (p. 182).

More recent reactions to digital disruption of traditional argument structure similarly exhibit a range from ambivalence to acceptance. Jim Kalmbach's (2006) admirable tribute to the nonlinearity enabled by hypertext, a review of 10 years of *Kairos* archives, characterizes his effort to corral the journal texts' various organizational structures as "exhausting" (Sorting the Webtexts section, para. 1). He lauds what he sees as an increased presence of an intuitive organizational structure in the *Kairos* Webtexts, one informed by a more standardized use of design principles, such as typographic contrast to indicate heading levels (*Kairos* and the Ransom . . . section, para. 3, 5). Meanwhile, Elizabeth Hayes's (2003) review of the impact of current-traditional rhetoric upon text generation exemplifies thoughts on arrangement as interpreted via social constructivism. Hayes argues that an emphasis on predictable form and structure is in constant tension with writers' (especially writing teachers') desire for freer text composition processes. Citing Fish (1995), Hayes argues that as more fields of study experience "the interpretive turn" and acknowledge that disciplinary value systems are discursive constructions, prescribed and replicable models for organizing text will lose relevance (The Contemporary Situation section, para. 16–17).

Further, the literature on issues of arrangement includes enthusiastic discussions of the spatial and linear thinking that can occur during composition, and how emerging and traditional text forms can be seen as representing these thought processes. Maryann Kope's (2006) comparison of "information literacy," "academic literacy," and "computer literacy," for example, states that digital fluency will include the ability to "encapsulate higher-level thinking in the context of information technology" (e.g., cognition of how the unique reading patterns of print and of digital texts in turn affect interpretation of content) and to "organize and navigate information structures" (pp. 75–76). She defines this type of literacy as a hybrid, a compact of skills considered part of traditional academic literacy—especially critical thinking, which she conceptualizes as the ability to perceive significant ideas from a digital text (pp. 71, 76).

In their study of the relationships among spatial and linear thinking and digital and print forms, Rainer Bromme and Elmar Stahl (2005) suggest that the reading and writing of digital texts are most often understood in terms of either "book" or "space" metaphors. The authors explain how writers who understand hypertext development via a "book" metaphor privilege issues of predictable chronology, often at the expense of the complexity with which ideas can be interwoven; meanwhile, those writers who compose with a "space" metaphor, whereby audiences can move/click per their unique preferences around pieces of textual content, are freed from the obligation of linearity and consequently are able to explore more fully the connections among developing ideas and to represent these connections more richly (pp. 117–118).

Citing Bereiter and Scardamalia (1987), Bromme and Stahl articulate how the interaction between writing acts and the construction of knowledge is most productive when a writer remains simultaneously aware of the information discovered through the research process *and* of the rhetorical exigencies of the situation (p. 116). Such mindful writers consider their researched information and construct their texts in austere keeping with the anticipated publication, such as a scholarly article for a peer-reviewed journal. Bromme and Stahl point out that difficulties immediately occur, however, when attempts are made to import this rigidly defined process into hypertext construction, most notably in the areas of structuring the content (p. 116). Even indicators such as the authors' reliance upon the hypertext-specific term *nodes* (perhaps not accidentally, the identical terminology used during the earliest popular discussions of hypertext [see Nielson, 1987]) signal the difficulties of attempting to distinguish between and then synthesize print and digital paradigms.

Bromme and Stahl argue that a "space metaphor seems to be more suitable than the book metaphor for fostering the interaction between the structures of the content and the hypertext required by the knowledge-transforming approach: It permits a correspondence between complex semantic structures and the text format" (p. 130). In their analysis of the efficacy of this metaphor, it is evident how disrupted hypertextual content may more closely approximate, and therefore better convey, nonlinear moments of knowledge transformation during the composing process. That Bromme and Stahl must dedicate extensive time to argue the space metaphor in lieu of the book metaphor speaks to the enduring influence of traditional structure whenever scholars assess digitized *dispositio*.

REACTIONS TO DIGITIZED *DISPOSITIO*: IMPLICATIONS FOR COMPREHENSION

Reactions to digitized text and its arrangement may differ in tenor, but this range suggests among other things the digital's potential to generate innovative ways of viewing the relations among thinking, writing, and persuading. As technology-mediated composing processes continue to appear and to fragment communications, scholarly description focuses accordingly on the features of digital text that influence comprehension.

By arguing that hypertext shards should be consumable as individual experiences, and so "satisfying" that they persuade the reader to follow the texts' hyperlinks to other shards, Landow (2004) works to reconcile readers' historical need for explicit structure with the fragmented arrangements that can appear in digital forms (Individual Lexias section, para. 1). His discussion is based on the premise that hypertext on so many levels is a novelty, both in experience and in entity, as the reader's ability to choose which links to follow is a qualitatively different experience from that of reading print—the very act of "following the link produce[s] surprise and delight" (The Pleasures section, para. 1). Landow argues that a digital writer's creation of "good" hypertext therefore should include a certain number of hyperlinks. He also states that those technological features that can be unnerving to the reader, such as including "multiple lines of organization," are not necessarily appropriate for every textual form that happens to be digital (The Document Should Exemplify section, para. 1).

Chronicling the interrelation between digital literacies and the professionalization of English studies, Thomas Miller (2005), long an important contributor to discussions of textual form, includes in his first hypertext essay a reflective coda on the comparisons between his argument online and that which appeared in the traditionally published *Rhetorical Education in America.* Miller asks and answers, "Okay, so how is clicking on links really all that different from reading a footnote and following a reference to the library? internet-work [sic] and the experience of the intertext are obviously quite different. significantly [sic] different[,] and far more expansive relationships become possible" (Coda on Codex section, para. 2). As one example of such a possibility, the organizational guidance offered by a hyperlink's wording has migrated to influence other print forms' reading cues: According to Miller, books now "have more topical cues" (para. 2) as a result of the influence of "hot links" on readers' navigational behaviors.

Miller's question also is answered in David Farkas's (2005) comparison of print and digital text structures. His is a telling analysis not of how hyperlinks can approximate a print text's hierarchy of headings, but of the moments when this representation breaks down. A Web page's hyperlinks, for instance, are not bound to mimic the arrangement and subordination of ideas, as would be suggested by the headings on a print page. Instead, a link can just as easily move a user purposefully away from the page and to a text with entirely different thematic content (p. 11). Farkas proposes that these different linking methods in turn have created considerations such as the question of whether to write "link text that will enable users to disambiguate the links and make correct navigation choices without the support of overview information," for example, explicit major headings or forecasting statements (p. 22). One study of the use of technology-mediated writing as a classroom assignment specifies that linking options such as those explored by Farkas expand reader methods of comprehension and hence the number of audience interpretations; in the case of blogs, for instance, a text's perceived style "may be reflected in either the writing *or* the selection of links passed along to readers" (Downes, 2004, p. 18; my emphasis).

Also focusing on hyperlinks as representative of the digital's impact on arrangement and comprehension, Jay Gordon (2005) suggests that "exploring hyperlinks can also be an excellent way to explore . . . the general concept of intertextuality," in that a hyperlink is simultaneously a digital design element, a method for guiding the reader's journey within and across digital texts, and a component of the text itself (p. 50). The opportunity for learning lies just as much in the shortcomings associated with using hyperlinks to understand textual connections: Questions of authorship and reader friendliness arise in light of the extent to which a hypertext's links reflect its structure. For instance, while an element placed "higher up" on a printed page (e.g., a major heading) generally is intuited as representing "more general contextual information" (p. 57) connected to the page's content, a link positioned toward the top of a Web page instead may unexpectedly connect to specific details or even unrelated content. These and several other investigations into structural decentering suggest how changes in the approach to organizing text might in turn encourage new understandings of that text.

REACTIONS TO DIGITIZED *DISPOSITIO*: IMPLICATIONS FOR AUTHORITY

As this review of technology's impact on argument structure in text has suggested, lack of consensus over the merits of digitized texts illustrates how a fragmented, decentered, flattened, or simply different textual structure can be a dramatic experience for those accustomed to or comforted by conventional exposition. In some cases, textual content that is conceptually disjointed may be accused of incoherency, of being cobbled together. Such disciplinary responses to digitized writing can be seen as fueled by the perceived challenges posed by technology to acts of authorship, in addition to motivations suggested earlier in this chapter. Fittingly, authorial norms especially being challenged are described in *On Rhetoric*:

> Since rhetoric is concerned with making a judgment . . . it is necessary not only to look to . . . argument, that it may be demonstrative and persuasive . . . , but also [for the communicator] to *construct a view of himself* as a certain kind of person and to *prepare the judge*. (Aristotle, trans. 1991; my emphasis)

In that recent technologies such as Web 2.0 protocols seem not just to allow but to encourage fragmented texts (e.g., the Wikipedia *stub*, or a short bit of text that has yet to become a completed Wikipedia entry [Wikipedia, 2008, para. 1), the long-recognized norms may be becoming minimized whereby a communicator enacts, or a text gives nod to, the construct of a sole, consistent, cohesive authorial voice.

Further, emerging from the dialectic between conventional and digitized textual structure can be critique of received knowledge as a seamless body of information that has been authored singly. Critique of authority was raised in Moulthrop's (1991) analysis of what is considered by many information scientists to be the first hypertext network: Project Xanadu. In his description of early hypertext as a synthesis of old and new manners of argument making, Moulthrop explains that the structural implications of digitized text are unnerving because they get at issues of actual writing behaviors, as much as at intangible philosophies about text construction (para. 3). He points out that discussions about hypertextual restructuring often are afflicted by a contradiction: Enthusiastic calls for creating hypertext content that subsume if not dissolve the juggernaut concept of authorship, made simultaneously with demands for a protocol that enshrines authorship via citation in a digital text of conventional sources (para. 9–11).

Looking to other disciplines may enhance our understanding of technology's impact on authorship and suggest potential ways to honor a variety of knowledges. Here Moulthrop's reference to conversations in the information science field is especially pertinent. This field's response to decentered structure in digital texts, as briefly chronicled below, ultimately offers to those involved in writing studies understanding of how such structure can fragment grand narratives as well as the concept of monovocal authorship.

A microhistory of the information science field's discussions regarding hypertext's birthing indicates that discipline's attunement to issues of structure and received knowledge. This awareness surfaces in homages to engineer Vannevar Bush (whose development of an electronic information retrieval technology, the Memex, is widely credited as anticipating the World Wide Web [Stewart, 2006]). Information scientists' responses to Bush's work illustrate disciplinary uses of digital texts that acknowledge and value a range of knowledge sources. In their collection, *From Memex to Hypertext,* James Nyce and Paul Kahn (1991) explore Bush's understanding of what we recognize today as prototypical hypertext. The Memex, they write, "supported the explicit recordings of 'trails of association,' something Bush saw as an attribute of human memory . . . prone to fading in its non-mechanical form" (p. 122). His interest in a digital system that would mediate and perpetuate nonhierarchical knowledge is described in another chapter "as a way to put together all the disparate decisions of an organization. . . . [H]ypertext is . . . an infrastructure, national and international, . . . linking together myriad types of information for an enormous variety of audiences" (Meyrowitz, 1991, p. 288). As users navigate through an electronic system, they leave a "trail" recalling their unique knowledge and interests.

Bush and then Nyce and Kahn see the collection of these trails as a formidable data resource representing and simultaneously honoring the community of all individuals interacting with and through the system. Michael Buckland (1998) concurs that it is the act of voicing alternative historical accounts of the history of the digital age that facilitates development of a "community of interest," namely, the discipline of information science. Through the work of Buckland and other historiographers attuned to alternative narratives, "the nurturing of a community interested in the history of information systems has been consciously cultivated by a series of steps taken . . . to build a supportive infrastructure" that acknowledges the contributions, heretofore invisible, of peers and institutions by means of the recovery of crucial documents (Antecedents section, para. 2). One instance of such an infrastructure might be seen in Project Xanadu. With a focus on how users' technology-mediated interactions give rise to differing beliefs (or "agendas") regarding the generation, storage, and retrieval of knowledge, founder Ted Nelson created Xanadu to be a digitally contextualized knowledge collection whose structure emphasized its power to neutralize *doxa*. Nelson casts Xanadu as an "alternative paradigm for a computer universe, with its own alternative history of the computer field and alternative ideas of media, computer life, and the nature of connections" (1999, p. 2).

This disciplinary focus on digitized organizational structure and the knowledges it honors appeared at HyperTEXT '87, the first conference dedicated to the technology (Nielson, 1987, Definition section, para. 2). The wide acceptance of alternative knowledge sources promoted by Bush, Nyce and Kahn, Buckland, and Nelson, however, here appear to be in tension with a growing call for explicit hierarchy in digital texts, seen in comments offered by conference keynote speaker Frank Halasz. As he argues,

> The things which we can link to or from are called nodes, and the whole system will form a network of nodes interconnected with links. Links may be typed and/or have attributes, and they may be one or bi-directional. . . . We have no mechanisms for treating structures of nodes and links (subnets) as structures in their own rights. (Nielson, 1987, Definition section, para. 2; Classifying section, para. 11)

Halasz's own speech metaconfirms an emphasis on information arrangement by explicitly listing and categorizing his items for discussion; he even invokes a chronological hierarchy, as seen in Jakob Nielson's comment that Halasz points out how "items 1–3 on his list were the most important" (para. 17). However, since HyperTEXT '87, many discussions from information science have interrogated how such arrangements of a body of received knowledge may deny digitally-enabled multivocality. Buckland (2006), for instance, in keeping with the emphasis on alternative knowledge sources, even has suggested that much of Bush's influence over digital technology should be credited to chemist Emanuel Goldberg. Buckland's *Emanuel Goldberg and His Knowledge Machine* offers a historiography of writings on the illustrious chemist Goldberg's work in film technology and on the influence of this work over subsequent electronic inventions. For Buckland, the influence of Goldberg's contributions to information technologies has been marginalized unaccountably by historians: "How and why did the record of a famous man of considerable achievements come to be so thoroughly erased?" he asks. This erasure "illustrates how accidental received history can be" (pp. xii, 255).

Recent comments by Nelson (2007) can be offered as one answer to Buckland's question; Nelson explores how a body of knowledge gathers strength as it is redistributed among the users who receive it. In this accumulation, the offerings of invisible contributors to that knowledge (such as Goldberg, as Buckland argues) are homogenized. Citing the Internet as an example, Nelson complains that

> The World Wide Web was not what we were working toward, it was what we were trying to prevent. The Web displaced our principled model with something far more raw, chaotic and short-sighted. Its one-way breaking links glorified and fetishized as "websites" those very hierarchical directories from which we sought to free users. (p. 2)

Ultimately, this analysis of the work of Nyce and Kahn, Buckland, and Nelson, as they represent voices from the information science field, emphasizes the merits of seeing digitally-mediated text structures as opportunities for critiquing received knowledge and univocal authority.

DIGITIZED *DISPOSITIO*: IMPLICATIONS FOR AUTHORITY IN WRITING/RHETORIC

That a digital text's structural navigability, as discussed by information scientists, can encourage deconstruction of grand narratives and the foregrounding of alternative and transgressive ways of knowing suggests how other scholars and teachers might use digital technology to negotiate issues of structure, authority, and readership. Michael Salvo (2007) is energized by the "electracy" of digital writing in its potential to widen conceptions of authority and in so doing to acknowledge a larger chorus of participants. Salvo bears out his point with a comparison of the differences between a recently composed print text and its subsequently constructed digital version. As he describes, the strategies typically associated with "standard argumentation—thesis, evidence, warrants, examples, lines of reasoning"—do not exist in the digital version. Instead, the digital content's users are immediately "immerse[d] . . . in the subject matter. Dropping you into our world, so to speak. There are arguments and examples here, certainly, but no overarching thesis trying them all together." In this sense, the digital text constitutes a freed, dynamic "remix" of the print text authors' claims (December 4, 2007, "Wither the Essay" thread, TechRhet).

Elsewhere, literature devoted to organizational issues in text (including Jason Farman's "Gertrude Stein in QuickTime," which appears in this collection) has turned its eye to the location of digital technology users to understand how decentered structure may unsettle authority. This focus may occur in tandem with, or perhaps in response to, the capacity for a Web site to minimize, homogenize, or downright constrain unique user needs and actions. During the early moments of hypertext's arrival in public and classroom spaces, Cynthia Selfe and Richard Selfe (1994) warned how alternative, fragmented textual arrangements as they exist digitally may deny multivocal and behavioral authority to alternative digital participants. The authors explained how digital writers may promote and connect ideas via interfaces that are paradigmatical, in turn constraining users from seeking out and connecting to different or even oppositional concepts. In other words, the structure of digital interfaces, their linkages and "mappings," might promote only sameness, an ideological homogeneity (p. 486).

Henry Jenkins and colleagues (2006) analyze how the digital age has encouraged a "participatory culture"—a technology-cultivated space allowing for an increased number and variety of knowledge sources, both creator and user. The authors elaborate pedagogical implications arising from the increase in digitized forms of information, including the need to assist student users in navigating content influx. Students in the digital age "must learn to recognize the relation between information coming at them from multiple directions and [the process of] making reasonable hypotheses and models based on partial, fragmented, or intermittent information" (p. 36). Jenkins et al. see *multitasking*, or "scanning for relevant shifts in the information flow while simultaneously taking in multiple stimuli" (p. 35), as one of the "core media literacy skills" (p. 22) necessitated by a participatory culture. Rather than indict multitasking as a form of distraction, these researchers acknowledge the importance of crediting a variety of information sources.

Gail Hawisher and Cynthia Selfe (2004) echo this validation of digitally supported authority. In their 350-plus interviews exploring users' methods for acquiring digital skills, the authors articulate a *mea culpa* concerning their teaching roles:

> Our professional radar is tuned so narrowly to the bandwidth of print and the alphabetic—to school-based and workplace writing—that we miss a great deal of the more interesting and engaging self-sponsored reading and composing students do on their own time. . . . And because we often miss such activities, our instructional practices and values, our interests, and the texts we read and compose may be moving further and further away from those that students consider important. (p. 676)

Additionally, scholars have considered the significance of *both* surfacing digital writers' multivocality and enabling digital users' authority (see Lipson & Day, 2005). Johnson-Eilola's (2001) analysis of online help systems includes a warning that technologically delivered information can imply transparency in the tradition of the Shannon-Weaver sender-receiver model; the content seems free of authorial bias and positions the user as passive recipient (pp. 120–121). Johnson-Eilola demonstrates how, when a digital text exists as a cohesive, identifiable genre, its content is separated out from contextualizing influences, truncating and simplifying the information. Creating digitized content, he argues, should involve resisting these too-basic implications. In doing so, digital writers can "help users understand communication, production, thinking, and living as an often messy, complicated, open-ended activity, one that often requires attention to not merely the simplest functional activities but also the larger frameworks and contexts of that work" (p. 126).

By warning that digitized textual structure can dictate the presence or absence of involved parties just as readily as print texts can, Anne Surma's (2005) *Public and Professional Writing* similarly argues the responsibility of digital authors to indicate a text's various influences. Surma focuses on the open forum organization of certain technology-mediated content—the assumption that "anyone" can post ideas to, say, a Web 2.0 interface such as Facebook—and its political consequences. Specifically, Surma advocates that the construction of Internet content be a democratic endeavor:

> Technologies of written language and the Internet *together* may be able to facilitate reflection, debate, discussion, decisions and further enquiry about the issues, needs, values and dreams of writers and readers. . . . In this view, one of the professional writer's key responsibilities is to ensure that the agency of individuals directly or indirectly addressed in this form of writing is not obscured, suppressed or ignored, but that that agency is, as far as possible, acknowledged or engaged. (p. 75)

In digitized *dispositio's* authorizing of multivocality, new possibilities are made available not just for composers and users but for larger disciplinary identities. David Kolb's likening of argument within digital structures to "Socrates in the Labyrinth" (1994) suggests that underlying the resistance to new lines of philosophical argument, here as present in hypertext, is the problematic assumption that "philosophy loses its identity if it is not structured as an argument" (pp. 324, 327). Kolb applauds the potential for breakdown during the movement of hypertextual information between author and reader, or the fact that "the author cannot control which links the reader will pursue" (p. 323). In his theorizing of the potential relations between crafting argument (here, as "do[ing] philosophy") and creating hypertext structure, Kolb makes note of how hypertext enables readers to appropriate authors' ideas and reassign them to new, perhaps even discordant, contexts. He further emphasizes that structural fragmentation by the digital, and the consequent diminished ability to retrace a putatively unbroken history of a received idea or body of knowledge, does not deny the philosopher's disciplinary purpose to create an "argumentative line" but rather complicates philosophical work by enabling the persistence of *multiple* lines. And allowing readers more choices in the routes they take to work out understanding enhances rather than weakens the power of philosophical argument (pp. 328–329).

In lauding the digital's ability to display the participation of more than one creator and to signal readers' various uses of a text, John Killoran (2005) delineates a "distinctly *Web* rhetoric" (p. 149, emphasis in original). Comparing how the designs of two official government Web sites situate users' discourses around each other in a manner that both honors and silences, Killoran shows how hyperlinks among a government organization's hypertext pages allowed conflicting political constituencies to actively "confront each other discursively"—what he views as an intertextual opportunity uniquely allowed by technology (p. 149). Salvo similarly posits that certain features of digital texts that "remix" voices and ideas, such as "essayistic . . . snippet[s]" (e.g., a short text on a page) and "near-to-hand" availability, make "snippet" composition and texts more favorable to writers and users. He believes that the disconnectedness, brevity, and other qualities encouraged by hypertext structure make writing a digital argument nonetheless an authoritative act. To Salvo, the rise of digital text production may mean that "scholarly literacy recedes into [even more of] a specialty," as such literacy historically has concerned more traditional print forms (December 4, 2007, "Wither the Essay" thread, TechRhet).

In these and other scholars' discussions of the digital rearrangement of knowledges, their authors, and users, the simultaneously spatial, messy, linear, and clear-sighted cognitive processes (such as those elicited by my efforts to create a nontraditional scholarly article) are being studied carefully. Structure, received information, authorship, and each user's unique rhetorical situation collude and collide via the digital, each encounter impossible to replicate. And as writers, writing teachers, and scholars increasingly reject the wholesale importation of traditional argument structure into digital forms, conceptualizations of the ethical, analytical, disciplinary, and political dimensions of digitized structure, and their inclusion of heterogeneous authors and ideas, will continue to evolve (Lamberti & Richards, n.d.).

CONCLUSION

When I began work on drafting a book review Webtext for *Kairos,* the project was not intended to generate an auxiliary investigation into discussions of argument structure and its divergences in digital forms, historical and current. Conventionally, the review genre would have led me to write an explicit thesis and a perceptibly linear argument structure that ultimately summed the book's assets and deficits. But in my struggle to organize the review was an unexpected opportunity, a chance to investigate divergent forms of *dispositio* presumably enabled by technology. More to the point, it offered me a chance to mull the key issues at stake and the forces behind my struggle—assorted writing interests who advocate their respective structural authorities.

Although Thomas Kent (1993) has argued that "nothing codifiable exists . . . that will ensure a correct interpretation of our utterances" (p. 88), the influence of "correctness," "norming," "standardization," et cetera nonetheless indisputably plays a role in communication, and in a textual sense can appear in figures such as grammar rules, generic conventions, and argument structure. Given the powerful force of discourse norming, only fairly recently has it been a scholarly option to reflect on the process of composing a hypertext review of a print book designed to resemble a Web site. It must be added that this chapter's organization, too, acts as an instance of the question of structure, in that it attempts to categorically review conversations from the literature regarding a largely nonlinear and dynamic phenomenon—the digital. The endeavor to "acceptably" classify content that frequently advocates nontraditional ideas joins similar efforts in the literature, such as James Suchan's (2006) reflections and analyses that he "dresse[s] in the genre of an academic research article" (p. 43). That this chapter's organizational approach could be called both appropriate and inappropriate to its content exemplifies how the issues regarding technological structure are complex indeed.

REFERENCES

Aristotle. (1970). *Metaphysics*. (W. D. Ross, Trans.). Retrieved March 20, 2008, from http://people.bu.edu/wwildman/WeirdWildWeb/courses/wphil/readings/wphil_rdg05_metaphysics_entire.htm

Aristotle. (1991). *On rhetoric: A theory of civic discourse*. (G. A. Kennedy, Trans.). New York: Oxford University Press.

Belling, L. (2007). *About writer's dreamtools*. Retrieved December 13, 2007, from http://www.writersdreamtools.com/view/about/default.asp

Bernhardt, S. (1993). The shape of text to come: The texture of print on screen. *College Composition and Communication, 44,* 151–175.

Bolter, J. D. (1991). *Writing space: The computer, hypertext, and the history of writing*. Hillsdale, NJ: Lawrence Erlbaum.

Bromme, R., & Stahl, E. (2005). Is a hypertext a book or a space? The impact of different introductory metaphors on hypertext construction. *Computers & Education, 44,* 115–133.

Buckland, M. (1998, October 23–25). *Overview of the history of science information systems*. Plenary session paper presented at the Chemical Heritage Foundation/American Society for Information Science Conference on History and Heritage of Science Information Systems, Pittsburgh. Retrieved January 14, 2007, from http://www.ischool.berkeley.edu/~buckland/chfintr.htm

Buckland, M. (2006). *Emanuel Goldberg and his knowledge machine: Information, invention, and political forces*. Westport, CT: Libraries Unlimited.

Christensen, C. M. (1997). *The innovator's dilemma*. Boston, MA: Harvard Business School Press.

Computer Chronicles. (Producer). (1987). *Hypercard* [video]. Retrieved December 13, 2007, from http://www.archive.org/details/CC501_hypercard

Delagrange, S. H. (2009). Wunderkammer, Cornell, and the visual canon of arrangement. *Kairos, 13*(2). Retrieved November 18, 2009, from http://kairos.technorhetoric.net/13.2/topoi/delagrange

Delany, P., & Landow, G. P. (1991). *Hypermedia and literary studies*. Cambridge, MA: MIT Press.

Dias, P., Freedman, A., Medway, P., & Pare, A. (1999). *Worlds apart: Acting and writing in academic and workplace contexts*. Hillsdale, NJ: Lawrence Erlbaum.

Downes, S. (2004, September/October). Educational blogging. *EDUCAUSE Review*, 14–26.

Eiler, M. A. (1998). *Nostalgic angels: A review*. Retrieved November 18, 2007, from http://kairos.technorhetoric.net/3.1/binder.html?reviews/eiler/nostal.html

Faber, B. (2007, October 22–24). *The problem of extraneous text: Opposition to organizational change, dynamic & synoptic orientations*. Paper presented at the 2007 SIGDOC '07, El Paso, TX.

Farkas, D. (2005). Explicit structure in print and on-screen documents. *Technical Communication Quarterly, 14*(1), 9–30.

Futura, R., & Stotts, D. (1990). A functional meta-structure for hypertext models and systems. *Electronic Publishing, 3*(4), 179–205.

Gordon, J. L. (2005). Teaching hypertext composition. *Technical Communication Quarterly, 14*(1), 49–72.

Gotlieb, I. (2004). Media investment management. *WPP Corporate Responsibility Report 2004*. Kent, UK: St. Ives Westerham Press, Ltd.

Hawisher, G. E., & Selfe, C. L. (2004). Becoming literate in the information age: Cultural ecologies and the literacies of technology. *College Composition and Communication, 55*(4), 642–692.

Hayes, E. J. (2003, November 8). *Alexander Bain's long shadow: The current-traditional paragraph in the classroom.* Paper presented at the Midwest Modern Language Association Meeting, Chicago. Retrieved December 15, 2007, from http://www.case.edu/affil/sce/Texts_2003/Hayesptr.htm

Jenkins, H., Purushotma, R., Clinton, K., Weigel, M., & Robison, A. J. (2006). *Confronting the challenges of participatory culture: Media education for the 21st century.* Chicago: John D. and Catherine T. MacArthur Foundation. Retrieved November 19, 2007, from http://www.digitallearning.macfound.org/atf/cf/%7B7E45C7E0–A3E0–4B89–AC9C–E807E1B0AE4E%7D/JENKINS_WHITE_PAPER.PDF

Johnson-Eilola, J. (1997). *Nostalgic angels: Rearticulating hypertext writing.* Norwood, NJ: Ablex Publishing.

Johnson-Eilola, J. (2001). Little machines: Understanding users understanding interfaces. *ACM Journal of Computer Documentation, 25*(4), 119–127.

Kairos. (2005). *Welcome to Kairos.* Retrieved November 17, 2007, from http://kairos.technorhetoric.net/

Kalmbach, J. (2006). *Reading the archives: Ten years on (nonlinear) Kairos history.* Retrieved November 18, 2007, from http://kairos.technorhetoric.net/11.1/binder.html?topoi/kalmbach/index.html

Kent, T. (1993). Formalism, social construction, and the problem of interpretive authority. In N. Blyler & C. Thralls (Eds.), *Professional communication: The social perspective* (pp. 79–91). Newbury Park, CA: Sage Communications.

Killoran, J. B. (2005). E pluribus unum? Dialogism and monologism in organizational Web discourse. *Journal of Technical Writing and Communication, 35*(2), 129–153.

Kolb, D. (1994). Socrates in the labyrinth. In G. P. Landow (Ed.), *Hyper/text/theory* (pp. 323–344). Baltimore, MD: The Johns Hopkins University Press.

Kope, M. (2006). Understanding e-literacy. In A. Martin & D. Madigan (Eds.), *Digital literacies for learning* (pp. 68–79). London: Facet Publishing.

Lamberti, A. P., & Richards, A. R. (n.d.). *Digital games, rhetoric, and democratic classrooms.* Manuscript submitted for publication.

Landow, G. (2004). *Is this hypertext any good? Evaluating quality in hypermedia.* Retrieved December 20, 2007, from http://www.dichtung–digital.com/2004/3/Landow/index.htm

Lipson, C., & Day, M. (2005). *Technical communication and the World Wide Web.* Mahwah, NJ: Lawrence Erlbaum.

Meyrowitz, N. (1991). Hypertext: Does it reduce cholesterol, too? In J. M. Nyce & P. Kahn (Eds.), *From memex to hypertext: Vannevar Bush and the mind's machine* (pp. 287–318). St. Louis, MO: Academic Press.

Miller, T. (2005). *Will English departments become the classics departments of the twenty-first century?* Retrieved December 9, 2007, from http://www.gened.arizona.edu/tmiller/rhetoric/englishdepartments.htm

Moulthrop, S. (1991). *You say you want a revolution? Hypertext and the laws of media. Postmodern Culture, 1*(3). Retrieved December 4, 2007, from http://www3.iath.virginia.edu/pmc/text–only/issue.591/moulthro.591

Nelson, T. H. (2007). *Ted Nelson.* Retrieved January 16, 2007, from http://ted.hyperland.com/

Nelson, T. H. (1999). Xanalogical structure, needed now more than ever: Parallel documents, deep links to content, deep versioning, and deep re-use. *ACM Computing Surveys, 31*(4es), 2–3.

Nielson, J. (1987). *Hypertext '87 trip report.* Retrieved November 18, 2007, from http://www.useit.com/papers/tripreports/ht87.html

Nyce, J. M., & Kahn, P. (Eds.). (1991). *From memex to hypertext: Vannevar Bush and the mind's machine.* Boston, MA: Academic Press.

Otlowski, M. (1998). The writing process and CALL: Hypermedia software for developing awareness of structure in writing. *Computer Assisted Language Learning, 11*(4), 419–425.

Ross-Larson, B. (2002). *Writing for the information age.* New York: W. W. Norton.

Salvo, M. J. (2007, December 4). *Re: Wither the essay* [Msg. 200712/72]. Message posted to the Association of Teachers of Technical Writing electronic mailing list, archived at http://lyris.acs.ttu.edu/cgi–bin/lyris.pl?enter=attw–l

Schilb, J. (2008). *Rhetorical refusals: Defying audiences' expectations.* Carbondale, IL: Southern Illinois University Press.

Selfe, C. L., & Selfe, R. J. (1994). The politics of the interface: Power and its exercise in electronic contact zones. *College Composition and Communication, 45*(4), 480–504.

Slatin, J. M. (1990). Reading hypertext: Order and coherence in a new medium. *College English, 52*(8), 870–883.

Stewart, W. (2006). *Vannevar Bush and memex.* Retrieved January 15, 2007, from http://www.livinginternet.com/i/ii_bush.htm

Suchan, J. (2006). Changing organizational communication practices and norms: A framework. *Technical Communication Quarterly, 20*(1), 5–47.

Surma, A. (2005). *Public and professional writing: Ethics, imagination and rhetoric.* New York: Palgrave Macmillan.

Swales. J. M. (1990). *Genre analysis: English in academic and research settings.* Cambridge, MA: Cambridge University Press.

Warnick, B. (2005). Looking to the future: Electronic texts and the deepening interface. *Technical Communication Quarterly, 14*(3), 327–333.

Wikipedia. (2008). *Stub.* Retrieved May 10, 2008, from http://gv.wikipedia.org/wiki/Wikipedia:Stub

Writing for the Information Age. (n.d.). Retrieved December 19, 2007, from http://www.wwnorton.com/catalog/fall02/004786.htm

Zappen, J. P. (2005). Digital rhetoric: Toward an integrated theory. *Technical Communication Quarterly, 14*(3), 319–325.

http://dx.doi.org/10.2190/CWDC3

CHAPTER 3

The Fourth Estate in an Era of Digitally Mediated Democracy*

Leonard Witt

OVERVIEW

In light of the lessons learned from past attempts in a variety of open-source endeavors, this chapter builds on open-source/open-content literature and applications. The chapter develops a framework in the form of an annotated checklist with which academics, citizens, critics, journalists, and the news media can explore models for reinventing traditional journalism. The chapter demonstrates how digital divergences exist between sustained corporate models and disruptive technology-generated models, between traditional journalism practices and citizen media breakthroughs, between proprietary and open-source practices, and between those who have done their research and those who have not. The extent to which these divergences are thought through and responded to will have profound effects on the future of journalism.

In 1998, Eric Raymond published "The Cathedral and the Bazaar," an essay exploring the potential of the open-source software movement. Raymond described how programmer Linus Torvalds was able, with a brief online request for help, to recruit volunteers from around the world who were willing to develop Linux from a kernel of code into a robust software operating system that now competes with the likes of proprietary-built Microsoft. According to the Open Source Initiative (2007), which strives for a unified definition of the term, *open-source software* must meet ten conditions that guarantee the right to use and to adapt the software without charge so long as others are permitted to build on modifications and derived works freely. Open-source software applications have become so prevalent and trustworthy that the open-source philosophy has begun to creep into other areas of activity in which

*This chapter is based on an article published in the journal *First Monday* (Witt, 2006a).

Killingsworth, M. J. (2005). From environmental rhetoric to ecocomposition and ecopoetics: Finding a place for professional communication. *Technical Communication Quarterly, 14*(4), 359–373.

Klein, N. (2002). *Fences and windows: Dispatches from the front lines of the globalization debates*. Toronto, ON: Vintage.

Martin, R. P. (2003). *Putting meat on the table: Industrial farm animal production in America*. Pew Commission on Industrial Farm Animal Production (pp. vi–ix). Retrieved March 4, 2010, from http://www.pewtrusts.org/uploadedFiles/wwwpewtrustsorg/Reports/Industrial_Agriculture/PCIFAP_FINAL.pdf

Owens, D. (2000). *Composition and sustainability: Teaching for a threatened generation*. Urbana, IL: NCTE.

Raisinghani, M. S. (2007). *Global information technology management in the digital economy*. Hershey, PA: Idea Group, Inc.

Rheingold, H. (2002). *Smart mobs: The next social revolution*. Cambridge, MA: Basic Books.

Rushkoff, D. (1994). *Media virus!: Hidden agendas in popular culture*. New York: Ballantine Books.

Schade, M. (2005). *The wasting of rural New York state.* Citizens' Environmental Coalition and the Sierra Club white paper. Retrieved March 4, 2010, from http://newyork.sierraclub.org/conservation/agriculture/index.html

Schlosser, E. (2001). *Fast food nation: The dark side of the all-American meal.* Boston, MA: Houghton Mifflin.

Scott, T. (2006). Writing work, technology, and pedagogy in the era of late capitalism. *Computers and Composition, 23*(2), 228–243.

Selfe, C. L. (1999). Technology and literacy: A story about the perils of not paying attention. *College Composition and Communication, 50*(3), 411–436.

Shirky, C. (2008). *Here comes everybody: The power of organizing without organizations.* New York: Penguin Press.

Shor, I. (1999). What is critical literacy? *Journal for Pedagogy, Pluralism, and Praxis, 4*(1). Retrieved May 21, 2008, from http://www.lesley.edu/journals/jppp/4/shor.html

Shreve, J. (2001, August 2). See our film, join our cause. *Wired.* Retrieved March 20, 2008, from http://www.wired.com/news/digiwood/0,1412,45628,00.html

Starke-Meyerring, D., Duin, A. H., & Palvetzian, T. (2007). Global partnerships: Positioning technical communication programs in the context of globalization. *Technical Communication Quarterly, 16*(2), 139–174.

Third World Majority. (2002). *What is the digital divide?* Retrieved March 20, 2008, from http://www.cultureisaweapon.org/digitaldivide/mainframe.php

Turnley, M. (2007). Integrating critical approaches to technology and service-learning projects. *Technical Communication Quarterly, 16*(1), 103–123.

Wilhelm, A. G. (2004). *Digital nation: Toward an inclusive information society*. Cambridge, MA: MIT Press.

new, easy-to-use online technologies facilitate collaborative work often accomplished among strangers. Yochai Benkler (2006), a prominent commentator on economic and social networking, observes that "the networked environment makes possible a new modality of organizing production: Radically decentralized, collaborative, and nonproprietary; based on sharing resources and outputs among widely distributed, loosely connected individuals who cooperate with each other without relying on either market signals or managerial commands." He calls this modality "commons-based peer production" (p. 60).

Just how thoroughly open-source software philosophies or commons-based peer production will be adopted in the mass communications industry remains an unanswered question. One predictor of the likely success of open-source movements is an awareness of best practices in open-source writing, an understanding that differs among journalists. When the *Los Angeles Times* unveiled its Wikitorial, an open-source experiment in commons-based editorial writing, it was shut down within two days, as vandals busily filled it with obscene photos (Shepard, 2005). The editorial-page editors evidently did not understand how open-source peer production works (Jarvis, 2005), for the Wikitorial had no defense against malicious or incompetent contributions. Even a cursory reading of the commons-based literature, such as Benkler's (2002) *Coase's Penguin,* would have predicted failure.

The move toward commons-based peer production in mass communications reflects the dual premises that news media audiences in the congregate know more than individual journalists do and that this form of production helps audiences become more involved in journalism projects or even in improving the quality and depth of journalism. As journalists consider commons-based peer production, known interchangeably as We Media, open-content or open-source journalism, participatory journalism, citizen journalism, and citizen media, they are well-advised to consider a framework built on the experiences of professionals in other fields. This chapter provides a framework both for applying these experiences and for understanding the divergences between traditional or mainstream journalistic practices and emerging possibilities for a reinvented, potentially more democratic, way of producing journalism.

FIRST, SOME CONTEXT

Clayton M. Christensen's (1997) phrase *disruptive technology* refers to technology that is "typically cheaper, simpler, smaller, and, frequently, more convenient to use" (pp. xii, xiv). For instance, the invention of the cell phone disrupted a long sustained model of business in land-line phone service. According to Christensen, "It is very difficult for a company whose cost structure is tailored to compete in high-end markets to be profitable at the low-end markets as well" (p. x). This is a profit lesson that the mainstream news media have learned only too well as they have been disrupted by the likes of Weblogs, Craigslist, Google, eBay, and Monster.com (American Press Institute, 2006).

However, the erosion of mass news-media dominance is not something that has occurred suddenly or recently. Phil Meyer (2005) has been warning the newspaper industry for years that it is losing audiences to niche markets and that younger audiences have been neither buying nor reading its products. (For an in-depth discussion on this topic, see his book, *The Vanishing Newspaper: Saving Journalism in the Information Age*); the erosion can be traced back as far as World War II. Nonetheless, the mass media, as local monopolies, remained enormously profitable until recently. With the average profit range hovering at around 20%, few were making major changes to confront the disruptions.

The message that mainstream media might well be a victim of disruptive technologies finally was heard loudly and clearly in late 2005. Within a 6-month period, daily circulation at newspapers across the country dropped by 2.6% (Seelye, 2005), and during that year more than 2,000 jobs in traditional news media were lost (Fitzgerald, 2005). So many layoffs were announced by major newspapers during one week in September that Steve Lovelady (2005), managing editor of the *Columbia Journalism Review*'s *CJR Daily* blog, wrote, "All in all, it has been a bloody week for newspaper journalism. And it's starting to feel as if the bow of the ship is slipping beneath the water" (para. 20).

In the midst of the Judith Miller/Valerie Plame/CIA outing coverage, for which *The New York Times* was being severely criticized in the blogosphere, Arthur Ochs Sulzberger, Jr., the *Time's* publisher, would tell critics that his stewardship would not be judged by singular incidents. "My job is simple," he stated. "My job is to work with . . . all of our colleagues and turn this great company, this institution, into something that will flourish in a digital age. I'm going to be judged on that. I'm not going to be judged on this story or that" (Auletta, 2005, p. 66). In short, Sulzberger believed that he would be judged mainly on how thoroughly he was able to respond to the challenge of disruptive technologies.

Members of the news media clearly were experiencing a significant shift in their attitudes toward the digital. Simon Waldman, director of digital publishing at *Guardian Newspapers* in the United Kingdom, observed that "We are getting the whole organisation ready for a digital future," predicting that newspaper operation in around 2010 would be 80% digital as compared with 20% in 2005 (Papers Prepare to Go Digital, 2005, p. 38). The public editor of *The Sacramento Bee* (2006) wrote, "With little fanfare, *The Bee* has embarked on the road to its future—a digital, multimedia, interactive, around-the-clock news universe. Everyone knows the final destination. It's the place where newspapers aren't on paper anymore but still provide readers with news and information, and make a profit, too" (Acuña, p. 3E). And Geneva Overholser (2006), then the Curtis B. Hurley Chair in Public Affairs Reporting at the Missouri School of Journalism, after the conclusion of a series of think-tank sessions with some of the top scholars and journalists in the United States, wrote in the first line of her call for action, "The long-building plaint is now undeniable: Journalism as we know it is over" (p. 5). Journalists such as Overholser and Sulzberger, the cardinals, the ultimate defenders of the cathedral of journalism, were declaring that the church was in ruins and a new one must be built.

NOW WHAT?

Economic pressures and loss of audience were persuasive factors helping news media leaders to recognize that business as usual was not working. At the same time, successful experiments in open-source journalism, citizen outlets such as South Korea's online *OhmyNews,* were drawing attention. Since the 2000 entrance of *OhmyNews*, which was the world's first online newspaper to publish reports by readers, more than 40,000 citizen journalists have written for the company, some having written more than 500 stories apiece (Min, 2005; *OhmyNews*, 2006; Yu, 2003).

Variations of *OhmyNews* are being experimented with around the world. Some, like *Northwest Voice* in Bakersfield, California, are adaptations affiliated with traditional newspapers; others, like *BaristanetNJ* in Essex County, New Jersey, and *H20town* in Watertown, Massachusetts, are tiny local startups serving their own small communities (*BaristanetNJ*, 2006; *H20town*, 2006; *Northwest Voice*, 2006). But although traditional news operations are standardized in terms of format and delivery system, no such formulas have evolved in the open-content, citizen-media movement. One site monitoring online journalism (*CyberJournalist.net*, 2006) counted almost 70 experimental approaches, but these were so fluid that it was difficult to keep an accurate count. Each, however, represented an effort to adjust to differences between predigital news gathering and distribution practices and rapidly evolving new practices and possibilities.

Trial and error is the norm, at times with less than stellar results. Dan Gillmor (2004), who "wrote the book" on citizen journalism (*We the Media: Grassroots Journalism for the People, by the People*), experienced difficulty himself in attempting to reinvent journalism: *Bayosphere*, his citizen journalism experiment, faltered within one year. As he explained, "Citizen journalists need and deserve active collaboration and assistance. They want some direction and a framework, including a clear understanding of what the site's purpose is and what tasks are required. (I didn't do nearly a good enough job in this area)" (cited in Witt, 2006b). Just as with the *Los Angeles Times*'s Wikitorial, *Bayosphere*'s poor showing could have been predicted by a careful reading of the open-source software and commons-based peer production literature. Beneath what might look like an open environment, there usually is structure and someone with a vision actively providing guidance.

Indeed, the uneven success of open-source journalism points to a divergence in knowledge bases among those who have done their research and those who have not. The open-source and commons-based peer-production literature provides clues for acquiring the knowledge needed to undertake projects in open-source journalism. This chapter provides a checklist based on these clues to assist individuals, groups, and institutions in reinventing mass communication, and specifically journalism, in the digital age. Although key recommendations are listed in numerical order, the recommendations are amorphous and nonlinear and thus can be blended and reordered to suit the case. As will be illustrated, one person trying to arrive at a big idea is usually no match for collective thought. Thus, the following is offered as Release 1.0, a kernel of an idea, which can and should be expanded on.

Step 1: Have an Attitude Check

Traditional journalists must be open to the change represented by citizen media breakthroughs. The success of many citizen media experiments depends on getting audiences involved; that is, on moving from a top-down to a conversational model. The spirit of these experiments is captured by Gillmor (2005) when he writes, "Instead of lecturing our audiences, let's ask for their help and offer ours. We can do great things together, and we should" (p. 13).

Making such a change will require a major shift in attitude for many journalists. Indeed, getting substantial newsroom cooperation will be no easy task, for newsrooms, according to a Readership Institute (2000a) study, operate largely in passive/defensive or aggressive/defensive modes (pp. 3–4). The report adds that the former mode "often leads these organizations to preserve the status quo rather than look for major opportunities and improvements" (2000b, p. 3). In the latter, "an unrelenting pressure to appear perfect and expert mitigates against customer service, admission of errors, trying new and perhaps risky things and teamwork" (p. 4). But this is a time when journalists must see the larger picture. As Kevin Kelly (2005) writes of the misjudgments of media experts writing on the early appearance of new media in journalism:

> Everything media experts knew about audiences—and they knew a lot—confirmed the focus group belief that audiences would never get off their butts and start making their own entertainment. Everyone knew writing and reading were dead; music was too much trouble to make when you could sit back and listen; video production was simply out of reach of amateurs. Blogs and other participant media would never happen, or if they happened they would not draw an audience, or if they drew an audience they would not matter. What a shock, then, to witness the near-instantaneous rise of 50 million blogs, with a new one appearing every two seconds. (p. 3)

Newsroom skeptics also must understand the potential altruistic nature of audiences. Benkler (2004) reminds us that people often are willing to share their lives with people they will never meet. They carpool with complete strangers and share their computers for scientific investigations. "Anyone," Benkler writes, "who sits in a New York City playground can only marvel at the paradoxical phenomenon of Wall Street traders admonishing their children to 'share nicely,' and will appreciate our deep cultural commitment to sharing some of our private, rival possessions" (p. 305). The literature is rich with the indication that software programs that rival or are better than those produced by proprietary operations can be built because the number of these potential volunteers, "good" hackers, is so great (Benkler, 2002; Raymond, 1998). Behind this fact is a simple intuitive notion. Ten carpenters can build a house faster than one, but only if certain conditions are met. The workers need to be motivated, need a plan of action, and need certain skill sets. Take away any of those elements, and the end result could be another Wikitorial or Bayosphere.

Step 2: Draw an Ecosystem

The *Newspaper Next: Blueprint for Transformation* report published by the American Press Institute (2006) focuses primarily on how disruptive technologies can be used advantageously by newspapers. It provides what might seem like counterintuitive advice; that is, to aim innovations at nonconsumers rather than at the traditional core audience (p. 20). This advice implies a leap from long-sustained practices to new models that will address digital disruptions. Writing about open-source music, Peter Hanappe (2005) notes that it is important for communication researchers to address the full open-source "ecosystem," including all exchanges and interactions among authors, reviewers, audiences, producers, and distributors (p. 200).

Likewise, each news media company needs to determine its own ecosystem. Some newsrooms are already familiar with *civic mapping*, or the construction of a schematic drawing of their communities that not only reflects the powerful and famous, but also cuts deeply into the fabric of social classes and grassroots organizations. According to Jan Schaffer (2002), civic mapping is "a systematic way of getting to know your community, of building a database of people who know what's going on in your community—and, perhaps, of challenging some of the newsroom stereotypes and preconceived notions about parts of the community" (para. 7). In this vein, the open-source ecosystem might be put online as a visual representation for the public. Contributors might add new groups and subgroups, upload comments about groups, and even move groupings around to allow a more complete understanding of the ecosystem's dynamics. Beth Noveck (2005) argues that these visual representations, aided by technology already present in online community-building games, can encourage citizens to become involved in deliberative exercises (Me, Myself and Avatar section).

Step 3: Do a Philosophy and Mission Check (How Free is Free, How Open is Open?)

Every news operation will have its own culture, philosophy, and mission. In the past, some newsrooms opted for a public journalism approach to newsgathering, which meant that they worked with audiences whom they treated at times as citizens engaged in problem solving, not merely as consumers. Other newsrooms wanted to remain aloof professional gatekeepers who would decide what the news was and then report it and let the masses decipher and use it as they would. New technologies, however, have changed the dynamics as the public storms the gates and forces former gatekeepers to rethink their roles (Witt, 2004).

The open-source software movement grew out of competing philosophies, which have, as will be explored in greater depth, similarities with the philosophies emerging within the We Media, open-content movement. Marianne van den Boomen and Mirko Schäfer (2005) compare the philosophies of early open-software pioneers such as Richard Stallman, Eric Raymond, and Linus Torvalds. For Stallman, founder of the Free Software Foundation, "Free software is not about money, but about principles. The four principles of freedom include: The freedom to use, the freedom to redistribute copies (either gratis or for a fee), the freedom to modify, and the freedom to distribute modified versions without needing to ask for anyone's permission" (p. 44). van den Boomen and Schäfer observe that "Anyone who uses the term 'open source' instead of 'free software' is suspect in Stallman's eyes" (p. 46). In comparison, Raymond, another high-credibility hacker, "can be seen as Stallman's opponent in the open-source paradigm battle. His focus on open source is not ethical but financial. . . . [For Raymond,] it is necessary to get rid of the notion of 'free software,' as this is too ideological and anti-business" (p. 48). Finally there is Torvalds, who, beginning with a small kernel of programming code, persuaded volunteers from around the world to write the code for the successful Linux operating system. van den Boomen and Schäfer summarize:

> Stallman, the ethical purist, Raymond, the libertarian business-model organiser, and Torvalds, the pragmatic hacker and benevolent leader. All three are necessary to keep the open-source actor network alive and kicking, each enlisting his own allies (sometimes overlapping with the others), and each transforming and speaking for different flavours of open-source products, whether they be pure or hybrids of "free" and "non-free" software. (p. 50)

In new open-content journalism paradigms, the first mission of those who want it all to be bottom-up, community-oriented journalism is to serve the public and *then* worry about the profits. Others demand that financial stability take precedence and the other pieces be addressed later. Still others seek a pragmatic mix of both objectives. The British Broadcasting Corporation (BBC) is already moving toward a freer approach, coming closer to Stallman's four principles of free software. In December 2005, it began a trial experiment by opening its archives for the first time for the UK public. According to the Association of Moving Image Archivists, visitors to the BBC site were "welcome to download the clips, watch them, and use them to create something unique" (2005, para. 3).

In the United States, the Public Broadcasting Service's (PBS) Digital Future Initiative Panel's report (2005) advises PBS to open its extensive archives as well. However, Jacoba "Coby" Atlas, as PBS' co-chief program executive, argued that "there are many obstacles . . . that are often not in our control or our producers' control—including copyright issues, Guild issues, third-party rights issues, etc. All of this has to be sorted out before you can provide content anywhere, anytime for free" (Witt, 2005, para. 21).

Of course, many perceive it as more reasonable for a nonprofit to give away its content than for a for-profit entity to do so. Thanks to the work of people like Lawrence Lessig in the "copyleft" movement, an individual or an organization has choices in this regard. For example, according to its Web site, the Creative Commons (2006) "offers a flexible range of protections and freedoms for authors and artists. We have built upon the 'all rights reserved' of traditional copyright to create a voluntary 'some rights reserved' copyright." In the digital era, each news organization must decide how much it wants to keep proprietary and how much it wants to make fully available to the general public or, at least, to its open-source contributors so they can remix or build upon the content for their own uses or for the use of their employer. These divergences in approaches to how open and how free a news entity should be in light of the digital are at the very heart of reinventing journalism.

Step 4: Come to Terms with Objectivity and Other Ethics of Professional Journalism

Contemplating his long career in communication, James Carey (1997) recalls that nearly 40 years earlier he had observed that in the latter part of the 19th century, reporters were trained to be "objective," and in this capacity used

> not intellectual skill as critics, interpreters, and contemporary historians but technical skill at writing, a capacity to translate the specialized language and purposes of government, science, art, medicine, finance into an idiom that can be understood by broader, more amorphous, less educated audiences. (p. 137)

Although journalists have long received a technical education that stresses objectivity, in practice,

> reporters mediate between the audience and sources, [and] they are pulled in two directions: Serving the interests of the sources or the interests of the audience, which are rarely identical. Often, if not usually, reporters develop contempt for both parties they serve: The audience because it is so often apathetic and uninterested, the source because it is so often dishonest. (p. 139)

Carey concludes that since traditional reporters are, in the end, closer to their sources than to their audiences, they turn their sources into their "ultimate audience" and became more "allied structurally if not sympathetically with the persons and institutions" about which they report (p. 139).

Carey's expanded arguments were among the seeds of the public journalism movement (Rosen, 1999, pp. 67–71), which asks reporters to work more closely with their audiences and to dig deeper into communities (Public Journalism Network, 2003, para. 6). The movement has been controversial, but this chapter is not the place to revisit arguments surrounding its conception. It is still appropriate, however, to remind professional writers that growing close to one's audiences always affects the dynamics of communication. The consequences, good and ill, of the new relationships brought about through public journalism are just one more divergence between past and present that writers must consider.

The controversy intensifies when citizens start producing their own forms of journalism, and, in effect, breach the dividing line that traditionally has kept journalists on one side of the newsmaking world, and audiences on the other. As Samuel Freedman (2006) writes, "To treat an amateur as [someone who is as] credible as a professional, to congratulate the wannabe with the title 'journalist,' is only to further erode the line between raw material and finished product" (para. 9). Yet, as Gillmor (2006) rebuts, "Citizen journalism and professional journalism are not mutually exclusive concepts. We're actually heading toward an ecosystem that will support a variety of journalistic endeavours" (Enormous Opportunity section, para. 2).

Should citizens be treated as journalists or, as Freedman advocates, as producers of raw material that later may be turned into finished journalistic projects? Or will the citizen journalists provide the bulk of the news as they do at *OhmyNews* in South Korea? Or, as not to confuse the lines between professional and amateur, should independent subsidiaries such as the *Bakersfield Californian's* professional/amateur-written *Northwest Voice* be developed? What will the ethical code look like for a news organization that begins to lower its gates and allows an open-content model? What will its legal obligations be?

Step 5: Determine a Citizen Involvement Model That Fits Your Newsroom

In the past, professional journalists kept an intentional distance between themselves and their audiences. In fact, the public journalism movement of the 1990s aimed in part to help journalists reach out to the public, but journalists were reluctant to change. Now new technologies and declining readerships are encouraging journalists to work with their audiences and to narrow this long-standing divide.

One model for getting outsiders to help contribute to news media has been in place for decades. A reporter gets a story idea, does some preliminary research, and eventually gets on the phone and begins to interview people. Some interviewees may be part of the news audience, some reliable sources, and some complete strangers, but, as any reporter knows, most are willing to share time and information. Conventionally, none is paid (pocketbook journalism being considered unethical by reputable journalists), and the interviewees' motivations can differ widely. Citizens may share their time and expertise out of a sense of civic duty; public relations officers because it's part of their job; consultants or academics because it enhances their reputation; hobbyists out of their love for the subject matter; social activists out of passion for a cause; narcissists for the gratification of seeing themselves in the paper or on TV, or heard on the radio. Some audience members are motivated by the pleasure of developing Web content. In the free and open-source software literature (Ghosh, Glott, Krieger, & Robles, 2002, section 4; Lerner & Tirole, 2005, pp. 102–105; Rossi & Bonaccorsi, 2005, Introduction, Table 1) these motivational words recur: *Reputation*, *career benefits*, *fun*, *community*, and *recognition*.

Benkler's (2004) main premise is that "sharing nicely" is so prevalent that it should have a place in economic theory. And for many hackers, money is not a necessary element, although at times it does have a role to play. As Torvalds (1998) states in an interview about developers, "The first consideration for anybody should really be whether you'd like to do it even if you got nothing at all back. And if you answer 'yes' to that question, then it is probably a project you'd enjoy doing, and one where concerns like how much you'd get out of it personally really are rather secondary" (fifth section, para. 4).

In terms of Linux, it is important to remember that two groups played key roles in the release of each version. First, program users discovered bugs or had opinions about how the program could be improved to meet their individual needs, and, subsequently, programmers rewrote the code to fix problems or to otherwise meet user needs. Often, but not always, users and programmers were one and the same. Thus, the involvement of users was crucial to the creation of a robust program. As Torvalds states, "The thing that makes 'real users' so interesting is that they have so different [sic] usage patterns from most developers, which is why a product that is solely targeted to developers usually lacks a certain stability and finish" (fourth section, para. 8).

Increasingly, users are "largely or completely doing without the services of manufacturers" (von Hippel, 2005, p. 2) as they find it feasible to "do it themselves" when it comes to information products. Individual users with just a little understanding of really simple syndication (RSS) feeds and other relatively unsophisticated technology or ready-made platforms can build their own customized online news pages. News operations can ignore what these individual innovators are doing, can solicit ideas for product improvement from them, or can provide toolkits to help authors customize content. Other industry manufacturers are doing so with positive results, in part because active users are suggesting changes that meet the needs of real users. Eric von Hippel writes, "Given access to a user-developed prototype, manufacturers no longer need to understand users' needs very accurately and richly. Instead they have the much easier task of replicating the function of user prototypes that users have already demonstrated are responsive to their needs" (p. 109).

Enhanced collaboration between media users and outlets has the potential to make everyone's job easier and to produce a better product. In the past, reporters and editors were criticized for writing for other reporters and editors and not for the audience—in this instance, users. Working with users would serve everyone better. In fact, here is an example where digital possibilities may help narrow a longtime divide between professionals and their audiences if the attitudes of the former can change.

Step 6: Determine Citizen Contribution Extent and Effort (Sharing Nicely Meets Citizen Journalism)

In the past, many if not most journalists saw a clear distinction between themselves as the producers, and their audience members as the passive receivers of information. Today, however, any digitally literate person with the required tools can become a content provider. The unanswered question is how much audiences will produce.

There is little agreed-upon terminology in the We Media, open-source, open-content, citizen-journalism, citizen-media world. But in this new interactive sphere, where many news media are making the online transition from a one-to-many to a many-to-many model, I would suggest focusing on two overreaching categories. One is the open-content, citizen-journalism model, and the other is the sharing nicely model, a term I borrow from and which is discussed in great length by Benkler

(2004). These categories can overlap, but if you look at the open-source software literature, the really hard workers, the ones who have done the lion's share of the work—the programmers—are a fairly discrete group. One authoritative survey (Ghosh et. al., 2002, section 2.1) indicates that most open-source software-development projects (Noveck, 2005, p. 14, note 123) are worked on by a fairly small cadre. One might assume that, likewise, in the open-content world, the heavy lifting and time-consuming chores would be done by very small groups.

Many people, however, may be willing to help with lighter, less time-consuming tasks. As Benkler makes clear, almost everyone shares nicely. They give blood, help friends move, carpool, allow scientists to share their home computers when idle, give directions, share toys, cook elaborate meals for friends, and, I would add, spend time talking to the media. Given the right conditions, people share. One consideration, however, is how much is being asked of a person. A stranger asking directions to a bus stop will receive many more positive responses than one asking for a ride downtown. Although given the right conditions, either request could be fulfilled, the former has the greater chances.

Benkler writes, "When a project of any size is broken up into little pieces, each of which can be performed by an individual in a short amount of time, the motivation to get any given individual to contribute need only be very small. This suggests that peer production will thrive where projects have three characteristics:"

1. They must be modular. That is, they must be divisible into components, or modules, each of which can be produced independently of the production of the others. This enables production to be incremental and asynchronous, pooling the efforts of different people, with different capabilities, who are available at different times.
2. For a peer production process to pool successfully a relatively large number of contributors, the modules should be predominately fine-grained, or small size. This allows the project to capture contributions from large numbers of contributors whose motivation levels will not sustain anything more than small efforts toward the project. . . . In addition, a project will likely be more efficient if it can accommodate variously sized contributions. Heterogeneous granularity will allow people with different levels of motivation to collaborate by making smaller—or larger-grained contributions consistent with their levels of motivation. . . .
3. [A] successful peer production enterprise must have low-cost integration, which includes both quality control over the modules and a mechanism for integrating the contributions into the finished product, [while defending] itself against incompetent or malicious contributors. (2002, pp. 378–381)

Benkler's notion of *heterogeneous granularity* might help members of the news media define their projects. Fine-grained contributions such as answering an occasional e-mail might be said to characterize sharing nicely. On the other hand, large-grained contributions, which are often required of citizen journalists, demand more time and greater motivation in terms of researching, reporting, and writing a story or producing one for broadcast.

Minnesota Public Radio's Public Insight Journalism project (2008) fits the more finely-grained, sharing nicely model. People volunteer to share information with reporters and editors on all sorts of topics related to their hobbies, jobs, war experiences, and such. With thousands of people already participating, the project has a deep pool from which to work, with each individual asked to make only a small commitment of time. The Public Insight Journalism Web page tells its audience, "About once a month, we'll ask you to share your observations, insights and experience. We then pass on your information to our reporters and editors who may follow up with a request for more information, or perhaps an interview. It's a great way to share what you know" (Share What You Know section, para. 1).

OhmyNews, a citizen journalism model that asks contributors to write stories, is much more labor intensive (Yu, 2003). The modular size is greater, as is the motivation to contribute. So a newsroom should decide upon the level of commitment it wants—and might receive—from its audience. Contributions could be fine-grained, as with the Public Insight Journalism project, or more granular, as with citizen journalism; or granularity might be heterogeneous.

Step 7: Decide Who Gets Invited and Who Gets Excluded

Journalists have always needed to be able to apply ethical principles in deciding who will be invited to participate and who will be excluded intentionally, and this reality must be considered in light of the power relations between professional and amateur news writers in the digital age, as well as between digital haves and have-nots. With online news, organizations also must decide whether to target audiences on Main Street or on Hong Kong's Shum Wan Pier Drive. Additionally, interest group boundaries can be a concern: Target dog lovers, cat lovers, or both? Some individual, group, or institution must make the decision regarding who gets to participate and who does not.

Because people can be rude (as the Wikitorial editors learned at the *Los Angeles Times* and the ombudsman learned at the *Washington Post* [Comments Turned Off, 2006], when online commentators posted a panoply of profanities), tolerance standards decided on either by the institution itself or by the whole community should be posted and enforced. There are many filtering possibilities—far too many to cover in total here, but Benkler provides four options for maintaining civility:

1. hierarchically managed review, as in the Linux open-software development;
2. peer review, as in Slashdot;
3. norms-based social organization, as in Wikipedia; and
4. aggregation and averaging of redundant contributions, as in Mars Clickworks.

So many people were involved in the last that malicious contributors and crackpots were identified easily because their input deviated too greatly from the norm (2002, pp. 441–443).

Step 8: Rethink the Issue of Timing

Divergences also exist between traditional and digital ways of timing story production. A key lesson from the free- and open-software movement is one of timing, or, in this instance, the decision about when to release a product for public consumption. Raymond (1998) writes that he had once

> believed that the most important software . . . needed to be built like cathedrals, carefully crafted by individual wizards or small bands of mages working in splendid isolation, with no beta to be released before its time. In this light, Torvalds's style of development—release early and often, delegate everything you can, be open to the point of promiscuity—comes as a surprise. No quiet, reverent cathedral-building here—rather, the Linux community seemed to resemble a great babbling bazaar of differing agendas and approaches out of which a coherent and stable system seemingly emerged only by a succession of miracles. That this bazaar-style seemed to work, and to work well, came as a distinct shock to many. (Cathedral section, para. 3–5)

Throughout most of the history of 20th-century journalism, no story was released before its time. The nightly news aired at an appointed hour or a newspaper that had been produced at an appointed time the night before arrived first thing in the morning. Now, news can be retrieved almost immediately, and with handheld devices it is retrievable not only any time, but any place.

Step 9: Be Transparent

Before the open-source software movement, as Raymond (1998) reminds us, the model for success was to toil secretly and then release the final, perfected product. Linux, on the other hand, is open for all to see and for many to alter. Today, media companies have to decide how much of their inner workings they want the public to see. CBS News, for instance, decided to institute its *Public Eye* (CBS) Weblog

> to bring transparency to the editorial operations of CBS News—transparency that is unprecedented for broadcast and online journalism. And what, exactly, is *transparency* in the evolving context of news reporting? There are several aspects, but put most simply, transparency refers to the fact that henceforth the journalists who make the important editorial decisions at CBS News and CBSNews.com will be asked to explain and answer questions about those decisions in a public forum. (para. 1–2)

Other examples of transparent news reporting facilitated by new media are *The Spokesman-Review*'s *Daily Briefing* Weblog, where "the editorial staff of *The Spokesman-Review* meets to discuss that morning's paper and the coverage planned for tomorrow. The blog involves readers in that process" (Irby, 2007, para. 15). Likewise, the *Wisconsin State Journal* invites readers to choose one of five possible stories for front page play the next day (Foley, 2006). If, in the past, audience access meant a letter to the editor or contacting an ombudsman, now each organization offers a much wider range of possibilities, as the model changes from lecture to conversation.

Step 10: Determine an Appropriate Ownership Model

Disruptive technologies force media organizations to look at the differences between mainstream, traditional forms of ownership and those that new technologies might impose. Although news ownership models run the gamut from private to public to nonprofit, it seems that some older models might need to be scrapped. At the very least, more competitive models need to be constructed.

Christensen (1997) writes that sometimes "creating an independent organization, with a cost structure honed to achieve profitability at the low margins characteristic of most disruptive technologies, is the only viable way for established firms to harness this principle" (p. xx). An excellent example of dealing with low margins is the citizen-written *Northwest Voice*, which was developed as an independent subsidiary of the larger, mainstream *Bakersfield Californian*. The alternative, citizen journalism, disruptive innovation project does not infringe on the traditional, sustained journalism model of its mainstream counterpart.

Furthermore, ownership structures and perceptions can affect levels of audience participation. Benkler (2002) writes that if people perceive themselves as gullible, they are apt to abandon their willingness to share nicely. He uses as an example the early AOL discussion board leaders, all volunteers who "left when they began to realize that their contributions were effectively going to increase the value of the company" (pp. 440–441). Indeed, some felt so aggrieved that they sued for back wages (Margonelli, 1999). However, as Benkler points out, the psychology of ownership is complicated: Why academics, he concludes, "are willing to accept the bizarre system in which they contribute to peer review journals for free, sometimes even paying a publication fee, and then have their institutions buy this work back from the printers at exorbitant rates remains a mystery" (2002, p. 441). John Killoran's chapter in this anthology helps explain this "mystery" by means of a framework acknowledging the value of cultural capital.

Step 11: Exercise Leadership

The digital era will highlight divergent styles of leadership. Traditional, top-down leadership must be reappraised to see if it can be made to fit open-source, collaborative paradigms. Explaining the advantage of the open-source paradigm whereby people only *seemingly* work for no return, Torvalds encourages us to "imagine ten people putting in one hour each, every day on the project. They put in one hour of work, but because they share the end results they get nine hours of 'other people's work' for free. It sounds unfair: Get nine hours of work for doing one hour. But it obviously is not" (1998, third section, para. 1).

Raymond (1998) elaborates:

> While cheap Internet was a necessary condition for the Linux model to evolve, I think it was not by itself a sufficient condition. Another vital factor was the

> development of a leadership style and set of cooperative customs that could allow developers to attract co-developers and get maximum leverage out of the medium. . . . But what is this leadership style and what are these customs? They cannot be based on power relationships—and even if they could be, leadership by coercion would not produce the results we see. (The Social Context section, para. 11)

Leaders in the open-source movement will foster trust and, using new technologies, enable others to work together on projects, often for no monetary gain, because they want to do so and not because they are required to do so.

Step 12: Decide Whether Reporters Will be "In Here" or "Out There"

In the early years of the digital era, divergences between the old brick-and-mortar workplace and a more fluid work environment were becoming marked. When Torvalds inspired people from around the world to develop the Linux code, he proved that Internet workers could produce great things without coming to the office every day—or ever. Likewise, each journalism organization will need to decide how important bricks and mortar are to its survival.

Yahoo! hired journalist Kevin Sites (see http://hotzone.yahoo.com) to travel globally, covering the world's hot spots. Everything he needs to be a multimedia reporter—a laptop, cell phone, and digital audio, video, and still cameras—fits in his backpack. He is a solo journalist, a SoJo, with no need for office space. Gannett's *News-Press* in Fort Myers, Florida (Marymont, 2006) is experimenting with mobile journalists, or MoJos, who spend half their time in the field showing citizens how to produce media. The new technologies mean that reporters such as MoJos or SoJos can file print, audio, and video content from anywhere. Newsrooms will have to decide if their staff are better stationed in newsroom offices or in places such as barbershops, community centers, and other places where their audiences and potential collaborators congregate.

Step 13: Experiment and Increase the Tolerance for Failure

Traditional media were for the most part risk adverse; this stance was often the result of a lack of competition. But as divergences between sustained models and disruptive models grow, a willingness to take significant risks may become necessary. Although much has been made of the phenomenal success of Linux, Beth Noveck (2005) makes clear that "it is easy to launch an open-source software development project where the collective mind of dozens of programmers [is] channeled into a common task, but most of those projects are never completed. In the end, those that do get done are largely the work of a handful of people" (Talk is Cheap section, para. 4). Moreover, von Hippel (2005) reminds us that false starts are not just experienced by the open-source software world, writing, "It is striking that most new products developed and introduced to the market by manufacturers are commercial failures" (p. 107). According to von Hippel, manufacturers that work with what he calls *lead users*, or those who modify a product to meet their own individual needs, will be better informed than manufacturers who rely solely on internally produced predictions.

This means going beyond traditional focus groups. To quote Kelly (2005) again, "Everything media experts knew about audiences—and they knew a lot—confirmed the focus group belief that audiences would never get off their butts and start making their own entertainment" (p. 3). But if experts had paid attention to lead users, or innovators, they would have had, in von Hippel's terms, a better understanding of what was coming in the world of commons-based peer production. Today we know that with their blogs, vlogs, and podcasts, there are countless lead users in the world of citizen-produced and -modified media. The news industry should be looking to these individuals for leadership as well as to their traditional audiences to increase the chances of successful experimentation.

Step 14: Create a Distribution System

Since the invention of the printing press, paper served as the mainstay of the printed word, and television of visual narrative; but a clear distance is developing between these older sustained models and new digital delivery systems. The days of three national television networks and monopoly newspapers are gone, or nearly so. Distribution platforms now include not only special interactive computer sites, but also mobile phones, personal digital assistants, portable media players, and all the other gadgets that audiences have or will have access to in the future. Yet with these new platforms come practical questions about design, information accessibility, and levels of interactivity. With interactivity come questions of filters for quality control, and perhaps automated trust and ranking systems.

In addition to practical questions about systems considerations, philosophical questions arise. "Far too little attention has been paid in the traditional literature to the impact of technology on the power of groups," writes Noveck (2005), who focuses "on the individual experience with tools rather than on the 'shared, social meanings'" (note 123). Observing that digital technology involves much more than simply e-mailing back and forth or chatting, she calls for the creation of programs that make it easy for participants to join in collective action, as might happen in face-to-face situations and as is discussed by Heeman Kim and William Faux in this collection. Should newsrooms facilitate such projects?

CONCLUSION

The 14 steps described in this chapter are guideposts, a starting place to help those interested in the evolution of communication in light of digital divergence to better cope with disruptions to the older models or, better still, to find ways of using disruptive innovations to develop innovative and effective ways of doing business. Of course, anyone who is paying attention will realize that the status quo is a moving target. Each day brings something new, and this, too, is a reality to adjust to.

There are, however, overriding lessons, as propounded by Benkler (2002) and Raymond (1998), whose essays have helped clarify the social and economic impacts of the open-source software and open-content movements. Each author stresses the relation between people and technology, a relation that will continue to grow in sophistication and power. According to Benkler, "Commons-based peer production presents a fascinating phenomenon that could allow us to tap substantially underutilized reserves of human creative effort. It is of central importance that we not squelch peer production, but that we create the institutional conditions needed for it to flourish" (p. 446).

Raymond's essay "The Cathedral and the Bazaar" helped many "nonbelievers" in the software industry see how people, often complete strangers, working together loosely in a "bazaar-like" atmosphere were able to produce a better product than that produced under hierarchy. Two of his ideas are well worth taking away as this chapter concludes. The first: "It's fairly clear that one cannot code from the ground up in bazaar style. One can test, debug, and improve in bazaar style, but it would be very hard to originate a project in bazaar mode. Linus (Torvalds) didn't try it. I didn't either. Your nascent developer community needs to have something runnable and testable to play with" (1998, Necessary Preconditions section, para. 2). In the case of journalism, it will be up to those who care about the future of their profession to produce "runnable and testable" ideas their open-content communities can play with.

Finally, also according to Raymond, "When you start community-building, what you need to be able to present is a plausible promise. Your program doesn't have to work particularly well. It can be crude, buggy, incomplete, and poorly documented. What it must not fail to do is convince potential co-developers that it can be evolved into something really neat in the foreseeable future" (para. 3). The crucial question for the news media is whether they can transform older, perhaps moribund, models into those that "can be evolved into something really neat in the foreseeable future." To do so, they must take what is of value from the traditional ways of thinking and doing and migrate with them into an era of uncertainty but great promise—the digital. There is no room for complacency.

REFERENCES

Acuña, A. (2006, January 29). The *Bee* journeys further into online news universe. *Sacramento Bee*, Metro Final, Forum, p. E3.

American Press Institute. (2006). *Blueprint for transformation*. Newspaper Next: The Transformation Project. Retrieved December 2, 2006, from http://www.newspapernext.org/2005/09/report_availability_1.htm

Association of Moving Image Archivists. (2005). *BBC open news archive*. Retrieved May 2, 2008, from http://palimpsest.stanford.edu/byform/mailing-lists/amia-l/2005/12/msg00190.html

Auletta, K. (2005, December 19). The inheritance: Can Arthur Sulzberger, Jr., save the *Times*—and himself? *New Yorker, 81*(41), 66.

BaristanetNJ. (2006). Retrieved January 20, 2006, from http://www.baristanetnj.com

Benkler, Y. (2002). Coase's penguin, or, Linux and the nature of the firm. *Yale Law Journal, 112*(3), 369–446.

Benkler, Y. (2004). Sharing nicely: On shareable goods and the emergence of sharing as a modality of economic production. *Yale Law Journal, 114*(2), 273–358.

Benkler, Y. (2006). *The wealth of networks: How social production transforms markets and freedom*. New Haven, CT: Yale University Press.

Carey, J. (1997). The communications revolution and the professional communicator. In E. S. Munson & C. A. Warren (Eds.), *James Carey: A critical reader* (pp. 128–143). Minneapolis: University of Minnesota Press.

Christensen, C. M. (1997). *The innovator's dilemma: When new technologies cause great firms to fail*. Boston, MA: Harvard Business School Press.

Columbia Broadcasting System. (2006). *About Public Eye*. Retrieved March 13, 2007, from http://www.cbsnews.com/stories/2005/08/30/publiceye/main805566.shtml

Comments turned off. (2006, January 19). *Washington Post*. Retrieved February 2, 2006, from http://blogs.washingtonpost.com/washpostblog/2006/01/shutting_off_co.html

Creative Commons. (2006). Retrieved December 2, 2006, from http://creativecommons.org/

CyberJournalist.net. (2006). *Citizens media initiatives list*. Retrieved January 28, 2006, from http://www.cyberjournalist.net/news/002226.php

Fitzgerald, M. (2005, December). The shrinking staff. *Editor & Publisher, 138*(12), 33–37.

Foley, R. J. (2006, January 29). Wisconsin paper lets its readers choose news. *Houston Chronicle*, Two Star Edition, p. 20A.

Freedman, S. (2006, March 31). *Outside voices: Samuel Freedman on the difference between the amateur and the pro*. CBS News's *Public Eye*. Retrieved April 8, 2006, from http://www.cbsnews.com/blogs/2006/03/30/publiceye/entry1458655.shtml

Ghosh, R. A., Glott, R., Krieger, B., & Robles, G. (2002, June). *Part 4: Survey of developers. Free/libre and open source software: Survey and study*. Retrieved February 1, 2006, from http:// www.infonomics.nl/FLOSS/report/Final4.htm

Gillmor, D. (2004). *We the media: Grassroots journalism by the people, for the people*. Sebastopol, CA: O'Reilly.

Gillmor, D. (2005, Winter). Where citizens and journalists intersect. *Nieman Reports, 59*(4), 11–13.

Gillmor, D. (2006, April 6). Why we are all reporters now. *BBC News*. Retrieved April 9, 2006, from http://news.bbc.co.uk/1/hi/technology/4882474.stm

H20town. (2006). Retrieved January 20, 2006, from http://h20town.com/

Hanappe, P. (2005). Building open ecosystems for collaborative creativity. In M. Wynants & J. Cornelis (Eds.), *How open is the future? Economic, social & cultural scenarios inspired by free & open-source software* (pp. 199–229). Brussels: VUB Brussels University Press.

Irby, J. (2007, January 31). Let me make this perfectly clear. *The Cavalier Daily*. Retrieved May 2, 2008, from http://www.cavalierdaily.com/CVArticle.asp?ID=29057&pid= 1537

Jarvis, J. (2005, June 21). Wiki cooties and the death of editorials. *BuzzMachine*. Retrieved January 20, 2006, from http://www.buzzmachine.com/archives/2005_06_21.html #009908

Kelly, K. (2005, August). We are the Web. *Wired, 13*(8). Retrieved January 29, 2006, from http://www.wired.com/wired/archive/13.08/tech.html

Lerner, J., & Tirole, J. (2005). The economics of technology sharing: Open source and beyond. *Journal of Economic Perspectives, 19*(2), 99–120.

Lovelady, S. (2005, September 23). The week of the long knives. *CJR Daily*. Retrieved January 20, 2006, from http://www.cjrdaily.org/behind_the_news/the_week_of_the_long_knives.php

Margonelli, L. (1999, October). Inside AOL's cyber-sweatshop. *Wired, 7*(10). Retrieved April 8, 2006, from http://www.wired.com/wired/archive/7.10/volunteers.html

Marymont, K. (2006, February 10). How they did it: Fort Myers' "Mojo" journalists search out news at the neighborhood level, identify community contributors. *News Watch.* Retrieved April 8, 2006, from http://www.gannett.com/go/newswatch/2006/february/nw0210-2.htm

Meyer, P. (2005, August 9). *A wake-up call: Can quality and trust save journalism?* Retrieved January 22, 2006, from http://www.restoringthetrust.org/day3.shtml

Min, J. K. (2005, Winter). Journalism as a conversation. *Nieman Reports, 59*(4), 16–19.

Minnesota Public Radio. (2008, May 4). *Public insight journalism.* Retrieved May 4, 2008, from http://minnesota.publicradio.org/publicinsightjournalism/

Northwest Voice. (2006). Retrieved January 20, 2006, from http://www.northwestvoice.com/home/

Noveck, B. S. (2005, November). A democracy of groups. *First Monday, 10*(11). Retrieved January 8, 2006, from http://firstmonday.org/issues/issue10_11/ noveck/

OhmyNews. (2006). *About OhmyNews.* Retrieved January 23, 2006, from http://english.ohmynews.com/english/eng_section.asp?article_class=8

Open Source Initiative. (2007, July). Retrieved May 2, 2008, from http://www.opensource.org/docs/osd

Overholser, G. (2006). *On behalf of journalism: A manifesto for change.* Retrieved December 1, 2006, from http://www.annenbergpublicpolicycenter.org/Overholser/20061011_JournStudy.pdf

Papers prepare to go digital. (2005, November 15). *The Australian.* Features, IT Today Computers, p. 38.

Public Broadcasting Service. (2005, December 15). Digital Future Initiative Panel. *Digital Future Initiative: Challenges and opportunities for public service media in the digital age.* Retrieved February 2, 2006, from http://www.newamerica.net/Download_Docs/pdfs/Doc_File_2766_1.pdf

Public Journalism Network. (2003, January 25). *A declaration for public journalism.* Retrieved February 19, 2006, from http://www.pjnet.org/charter.shtml

Raymond, E. S. (1998, March). The cathedral and the bazaar. *First Monday, 3*(3). Retrieved January 26, 2006, from http://www.firstmonday.org/issues/issue3_3/raymond/

Readership Institute: Inside newspaper culture. (2000a, July). Retrieved January 29, 2006, from http://www.readership.org/culture_management/culture/inside_culture.htm

Readership Institute. (2000b, July). *The five-minute guide to culture.* Retrieved December 1, 2006, from http://www.readership.org/culturemanagement/culture/data/five_minute_guide.pdf

Rosen, J. (1999). *What are journalists for?* New Haven, CT: Yale University Press.

Rossi, C., & Bonaccorsi, A. (2005, May). Intrinsic vs. extrinsic incentives in profit-oriented firms supplying open source products and services. *First Monday, 10*(5). Retrieved February 11, 2006, from at http://firstmonday.org/issues/issue10_5/rossi/

Schaffer, J. (2002, October 4). Civic mapping defines community: A primer. *American Society of Newspaper Editors.* Retrieved January 11, 2006, from http://www.asne.org/index.cfm?id=4067

Seelye, K. Q. (2005, November 8). Newspaper daily circulation down 2.6%. *The New York Times,* Late edition, Final, p. 8C.

Shepard, A. (2005, June 21). Postings of obscene photos end free-form editorial experiment. *The New York Times*, p. 8C.

Torvalds, L. (1998, March). FM interview with Linus Torvalds: What motivates free software developers? *First Monday, 3*(3). Retrieved February 11, 2006, from http://www.firstmonday.org/issues/issue3_3/torvalds/

van den Boomen, M., & Schäfer, M. T. (2005). Will the revolution be open-sourced? How open source travels through society. In M. Wynants & J. Cornelis (Eds.), *How open is the future? Economic, social & cultural scenarios inspired by free & open-source software* (pp. 31–67). Brussels: VUB Brussels University Press.

von Hippel, E. (2005). *Democratizing innovation*. Cambridge, MA: MIT Press.

Witt, L. (2004). Is public journalism morphing into the public's journalism? *National Civic Review, 93*(3), 49–57.

Witt, L. (2005). Renaming PBS: Will it be public service media? *PJNet—Public Journalism Network.* Retrieved January 30, 2006, from http://pjnet.org/weblogs/pjnettoday/archives/000937.html

Witt, L. (2006a). Editorial history: An earlier edition of "Constructing a Framework to Enable an Open Source Reinvention of Journalism." *First Monday, 11*(6). Retrieved March 4, 2010, from http://firstmonday.org/issues/issue11_6/witt/index.html

Witt, L. (2006b). Gillmor on why his citizen media project faltered. *PJNet—Public Journalism Network.* Retrieved May 5, 2008, from http://pjnet.org/post/ 829/

Yu, Y. (2003, September 17). *OhmyNews* makes every citizen a reporter. *Japan Media Review.* Retrieved January 28, 2006, from http://www.japanmediareview.com/japan/internet/1063672919.php

PART II

Shaping the Professions

http://dx.doi.org/10.2190/CWDC4

CHAPTER 4

Gertrude Stein in QuickTime: Documenting Performance in the Digital Age

Jason Farman

OVERVIEW

This chapter investigates an interactive CD-ROM and a subsequent three-disc DVD video document created by the New York performance troupe The Wooster Group, interactive documents with implications for reinscribing the role of the audience in relation to performance. Comparison of the CD-ROM to the Group's live performance as well as to Gertrude Stein's text *Doctor Faustus Lights the Lights* illustrates how a digital mode of production gives rise to tensions between textual authority and reader agency.

Documents have traditionally maintained a specific trajectory, that being from event to record. A public meeting, for example, may take place at city hall and subsequently be written up for the local newspaper; likewise, in the music world, a band creates music in a live setting and later records an album, documenting the songs as they might have sounded if performed in concert. Documents typically attempt to represent or to re-create the live event as "accurately" as possible since documents seen as distant from live events often are considered "secondhand," or subordinate to them. Thus, as Philip Auslander (1999) notes in his book *Liveness*, few rock records "foreground the artifice of their studio construction; most are made to sound like [live] performances that could have taken place, even if they really didn't (and couldn't)" (p. 64).

With the advent of digital technologies, the relation between event and document (and the document's subordination to the live event) clearly is being troubled. This is especially true of documentation in live theater. Enthusiasts of musical theater often purchase the CD or DVD version of a show long before they see it live. And encounters with documents of a performance create in audiences a specific relation to the document that informs the way they later react to the live event, during which any deviation from the CD or DVD (e.g., changes in lyrics, notes hit, tonal qualities of a singer's voice) is noted immediately and marked as "different from the CD." To this extent, the document is not relationally subordinate to the live; rather, the live is understood as referencing the document.

Digital documents also contribute to this reversal through their ability to involve the *spectator-user* (a term derivative of Augusto Boal's [1985] *spect-actor*, who is both a spectator of the theatrical piece and an agent in its interactive creation) by enabling him or her to alter a work's narrative path. Interactive alteration of narrative and visual structures by either shifting a story in a new direction or simply moving the mouse and scanning a QuickTime panorama is a feature that distinguishes digital from analogue documents. The interactive CD-ROM—in fact, any digital technology under the umbrella "interactive"—often is read as a medium that shifts authority away from the playwright/author and moves it into the hands of the spectator-user. Yet, as this chapter will explain in light of a discussion of the New York troupe The Wooster Group, the term *interactive* often is criticized by media theorists as nothing more than a reinscription of various modes of authority; for although the spectator-user is able to direct the narrative path, other individuals created the possible paths and coded the media to allow for interactivity.

One of the key questions that this chapter grapples with is whether digital documents offer greater opportunities for agency to the spectator-user than analogue documents do. Put another way, to what extent does the correspondence between event and document remain in place with digital documents, and to what extent might interactive documents allow for a new configuration of the relation between live event and document? For instance, can an interactive CD-ROM come "closer" to documenting a performance than a video recording does, perhaps by better representing the reality of a production by simulating the spectator's ability to look around the various parts of a stage at will? Additionally, the chapter considers the ways in which interactive interfaces inform the emerging roles of the audiences of artistic representations in the digital age.

The blurring lines between live and digitized events are a constant trope in the performances of The Wooster Group, who have made the incorporation of documentary technologies (cameras, televisions, recorded sounds, and movie clips) into live performance an essential part of their productions. In keeping with their goal of blending live event and digital document, they commissioned director Zoe Beloff to create an interactive CD-ROM document of their live performance of *House/Lights*, which was based on Gertrude Stein's *Doctor Faustus Lights the Lights* (1938). The CD-ROM, titled *Where Where There There Where* (now available online at www.zoebeloff.com/where), reformats Stein's print text into an interactive interface that enacts the tensions between live and digital. As Beloff (n.d.) states, she wanted the project to "provoke questions concerning how we think about our mental space within the real world in relation to the virtual machines which we more and more substitute in its place" (para. 7).

The decision for The Wooster Group to create an interactive CD-ROM document of their *House/Lights* performance (1998), which was followed by a three-disc DVD video document of the performance (2004), has several consequences in terms of the documentation of theater in the Digital Age and the reinscription of audience roles in performance. This chapter's comparison of the CD-ROM to the live performance as well as to the written text of *Doctor Faustus Lights the Lights* suggests how a digital mode of production gives rise to tensions between textual authority and reader agency and how readers can intervene in digital production of text. These tensions are troubled further when we consider differences between the CD-ROM and the DVD documents.

TURNING ON THE ELECTRIC LIGHTS

Though many studies examine the *form* of interactive art and performance (e.g., how the piece involves the audience, how interactivity is staged, how tools are used to trouble the relation between material and virtual space), the works I study here rely on an interdependence of content and form. Stein's work is famous for the intimate relation between these categories. Her *Doctor Faustus Lights the Lights* was scripted as an operatic libretto derived from many earlier Faust stories, including those of Christopher Marlowe (1604) and Johann Wolfgang von Goethe (1808). Though the play is founded on widely familiar narratives, the content and the form deviate significantly from previous stagings. Traditionally, the story follows the character of Doctor Faustus, who famously sells his soul to the devil (here, the character of Mephisto, or Mephistopheles) for power, knowledge, and fame. In Stein's retelling, the title character sells his soul for electric light. After the deal is made, he initiates several debates with Mephisto regarding the metaphysical status of the soul, arguing that he actually had no soul to sell. The debate continues throughout the piece and becomes as old and tiresome to Faustus as the electric lights do. Nothing seems to interest him any longer, including the presence of a woman under the dual name Marguerite Ida/Helena Annabel, who comes to Faustus seeking help after a viper has bitten her. Nothing has meaning to Faustus because everything around him seems to have exited the realm of the "real." Instead, it seems to him that all he comes in contact with may be a simulation of something real, just as the electric lights are. By the end of the play, as Marguerite Ida/Helena Annabel becomes more and more the central protagonist of the piece, Faustus begs Mephisto to take him to hell if such a place exists.

The form of the play is also not left to the traditional textual styles of Marlowe or Goethe. Stein's *Doctor Faustus Lights the Lights* contains no obvious stage directions, nor is it always clear which character is meant to be speaking. Stein offers up her text to various interpretations by eliminating the typical structure of a dramatic piece, removing most of the punctuation, and blending dialogue and stage direction. Stein explains that she was attempting to evoke concepts without referring to their linguistic signs, which come attached to the audiences' mental schema. She says, "I was trying to name objects without using their names" (Adato, 1970). Insofar as the dialogue in the text is disconnected from actors and their actions, her enigmatic use of language complicates the process of signification. Moments in the text collide and conflate—dialogue becomes stage direction, and vice versa. Stein's text becomes one in which oppositions are broken apart and questioned.

It thus functions perfectly as the source for performances that question the divide between event and digitally produced document, such as The Wooster Group's production of *House/Lights* (including the CD-ROM and DVD subsequently produced from this performance).

The Wooster Group and its director, Elizabeth LeCompte, premiered its interpretation of Stein's text in 1997 at the Wexner Center in Columbus, Ohio and brought it to New York's Performance Garage in October 1998. *House/Lights* is an amalgamation of *Doctor Faustus Lights the Lights* and a 1964 B-movie by Joseph Mawra called *Olga's House of Shame*. These two sources commingle and reflect each other throughout The Wooster Group's performance. *House/Lights* stages various characters—such as Faust, Marguerite Ida/Helena Annabel, and Mr. Viper— through a single actor, Kate Valk. She speaks the text for these characters through a "dictaphone-like receiver/amplifier" that modifies the pitch and tenor of her voice (Callens, 2002, p. 128). Through the use of video displayed on the television screen in front of her (one of four televisions on the set), Valk splits the lines of Marguerite Ida/Helena Annabel. She speaks several lines, after which a still shot of her face is displayed on the television screen in front of her. Valk then tilts her head to the side and continues speaking the lines of Marguerite Ida/Helena Annabel, while the still image and the live-feed image of her are simultaneously displayed on the screen. This technique, used throughout the production, allows Valk to embody several characters at once. Her "live" video image interacts with the recorded still video image frozen on the television screen (Figure 1). This technique simultaneously forces a multiplicity of characters to exist in digital form on the four televisions. By being located primarily on the screens (rather than on the material, live stage), characters are distanced physically from the actor's body. Thus the tension between the "live" and the "mediated" is foregrounded.

Figure 1. The *House/Lights* DVD. Pictured: Kate Valk. The split character of Marguerite Ida/Helena Annabel exists only on the stage-front television screen, as the live-feed image of actress Kate Valk is layered over a still image of her taken moments earlier. (© The Wooster Group)

The distance created here between actor and character and between digital space and material space is a technique often used in theatrical performances seeking to draw performers and audience members into an awareness of the mediated nature of production. The technique was made famous by playwright and director Bertolt Brecht, who termed this the "Alienation Effect" (1957). This style of performance is in direct contrast to that which seeks to immerse the audience through realistic representations of daily life. Instead, the Alienation Effect, achieved by distancing actors and audience from immersion, places demands on the audience to interact with the narrative of the piece by coming to understand its plot and characters as constructed elements. One example of The Wooster Group's application of this technique can be seen in its positioning of the computer console that runs the sound and lighting effects of the performance. A computer such as this one typically would have been hidden in the back of the theater but instead takes a prominent stage front location. This foregrounding of the technology, usually used to create an aura of verisimilitude, immediately creates a visual tension over who or what is the artistic and authorial agent. If the character of Marguerite Ida/Helena Annabel exists only through technological means, does the staged technology (and its technicians) have more control over the performance than the actors do? Johan Callens (2002) notes that Valk's role

> was visibly contested by Tanya Selvaratnam [another actor in the performance], whose almost permanent bodily presence behind a laptop computer, suggested a greater directorial control (over some video images, effects?) than that of the Macintosh's vocal intrusions with which she was credited in the program (animal-like sounds, the announcement of the intermission, the voice-over calling Faust, a countdown). Last but not least, when the computer flew off at a tangent, offering other "words that rhyme with *bright* with *light* with *might*" than ones a bored Faustus recited, any vested authorial control was further challenged by language and technology, as in Stein's motoric experiments. (p. 128)

Jim Findlay's set design places Selvaratnam and the computer stage front, next to the television screen on which we are presented with Valk's image (Figure 2). These technologies frame the entire performance, de-emphasizing "the presence of the performers by pulling our attention away from their physical bodies and toward their images on the video screens" (Gendrich, 1998, p. 380). The co-presence of Valk and Selvaratnam may lead the audience to wonder which takes the dominant role in the performance: The live body of the actor, or the computing technology that allows the mediation to occur. The important role that computers and other digital technologies play in The Wooster Group's stage performance carries over into the translation of *House/Lights* onto CD-ROM format.

TURNING ON THE COMPUTER

By constantly reorienting the focus away from live bodies toward their representations on screen, The Wooster Group highlights technology's role as the

Figure 2. The *House/Lights* DVD. Pictured (l–r): Kate Valk and Tanya Selvaratnam. The stage focus is pulled from Valk's live presence during the performance of *House/Lights* to the computer station on stage, in which Selvaratnam controls several of the stage elements of the performance. (© The Wooster Group)

metaphorical equivalent of Faustus's electric lights, as a simulator or representer of "the real." What is more, in the Wooster Group's interactive document of *House/Lights*, the video and computers that had been onstage objects become the stage itself. The interactive CD-ROM performance *Where Where There There Where* includes QuickTime virtual reality (VR) panorama interfaces through which spectator-users can scan 360 degrees, look up and down, and zoom in on objects in the field of vision. Spectator-users exercise agency primarily through interactions with these VR panoramas, as they can click on spots on the interface and then enter a new panorama. The movement between QuickTime panoramas changes the narrative and visual progression of the performance document.

The CD-ROM's beginning black and white panorama is an abandoned factory under a cloudy sky (Figures 3 and 4). Embedded in the panorama of the factory are actors carrying out simple actions that loop and repeat, such as Roy Faudree's continuously writing on a chalkboard. Also against the backdrop of the factory are clips from the Krazy Kat cartoons of the 1930s.

The scene of the next interface is similar to that of the previous one—a spectator-user can scan the scene and click on various spots. The spots clicked on by the spectator-user will either activate a new QuickTime VR panorama—with actors and cartoons acting within the theatrical space—or take the spectator-user to a short movie clip. These small QuickTime movies (160 × 120 pixels—the standard size for embedded QuickTime files at the time the CD-ROM was produced) display The Wooster Group cast repeating simple movements, often in relation to a movie

Figure 3. Panorama of the abandoned factory that spectator-users first encounter in the interactive CD-ROM *Where Where There There Where*. The spectator-user, through new media technologies, can scan this scene virtually and travel through the corridors of the abandoned factory. (© Zoe Beloff and The Wooster Group)

Figure 4. QuickTime movie frame from *Where Where There There Where*. The inside of the abandoned factory that spectator-users first encounter in the interactive CD-ROM. The spectator-user can pan 360 degrees and click on various spots. (© Zoe Beloff and The Wooster Group)

clip from Stein's era, such as the aforementioned Krazy Kat cartoons, war footage, films on electricity, a Soviet film on Ivan Pavlov, and lesbian erotica ("because Beloff [the director of the interactive CD-ROM] thought that Stein would enjoy 'some lovely lesbian pornography from the 30s'" [Kilpatrick, 2000, p. 72]).

Beloff is heard speaking the entire dialogue of the performance. She recites lines from *Doctor Faustus Lights the Lights* and often intermingles them with accounts of figures such as Ludwig Wittgenstein, Alan Turing (a key developer of computer coding and artificial intelligence), or Pavlov—referencing readings that operate outside the *Doctor Faustus* text, to comment on the human-machine relations explored in the text and the *House/Lights* performance.

The disconnected quality achieved through the spoken text, the actors' movements, and the Stein-era film clips (each highlighted in *House/Lights*) reinforces the aesthetic agenda of The Wooster Group's style of performance, an agenda that includes exploring the relation between live and digitally mediated performance. Bethany Haye (1984) identifies four prominent features of The Wooster Group's aesthetic, which can be seen in the examples of *House/Lights* and the interactive CD-ROM: Iconographic references to worlds created in earlier pieces, textual condensation, distancing of character and text through filtered dialogue or unique casting choices, and recontextualization through theatrical ambience or physical environments that distort and dislocate (p. 166). In its CD-ROM, The Wooster Group's use of Stein's *Doctor Faustus Lights the Lights* uses all of these features. *Where Where There There Where* stages Stein's text in a performance that creates a theatrical space appropriate to Stein's aesthetic, a space in which "the text is as important as the video image is as important as the sound, and nothing has dominance" (Kilpatrick, 2000, p. 73).

INTERACTIVE PERFORMANCE ON THE DIGITAL STAGE

An ability to interact with the CD-ROM by panning across QuickTime landscapes or clicking elsewhere seems initially to offer the spectator-user an authority unparalleled in other forms of documentation (e.g., video documents in which the camera angles and narrative paths remain static). Given the history of theatrical interactivity, such agency cannot be called new. The seeming advantage of using this form of media in documenting performance, however, is its ability to alter the traditional narrative structures of sequentially linear texts and to place the ability to do so in the hands of the spectator-user. As Beloff writes, "Unlike film or video, [The Wooster Group CD-ROM] is a work that is designed to be experienced more than once, since each time it is played, the user can choose a different route and discover new ways to interpret the characters" (n.d., para. 4). This medium differs greatly from other forms of theater that allow the audience to choose narrative paths—such as Ayn Rand's *Night of January 16* (1934), which allowed audience members to choose the ending of the play—in that it dislodges the necessity for a sequential narrative structure.

That the spectator-user can participate through such acts in disturbing the narrative sequence demonstrates how interactivity is a central component of emerging forms of digital documentation. To fully interrogate *interactivity*, though, it is helpful first to examine the initial promise of this concept to enable diverse narrative forms, including the scholarly enthusiasm accompanying this promise. The interactive CD-ROM has complicated the reasons for "the sequential development of ideas," as Sue-Ellen Case (1996) writes, in narratives developed out of a technologically focused culture rapidly becoming populated by a "click elsewhere" audience (p. 30). Sequentially dependent narratives, according to Case, connect politics and place via "patriarchal mechanisms" (p. 30). In a critique of linearity and "the play of vertical and horizontal as inscriptions of gendered, hetero-social constructs," Case quotes Luce Irigaray:

> We need to proceed in such a way that linear reading is no longer possible: That is, the retroactive impact of the end of each word, utterance, or sentence upon its beginning must be taken into consideration in order to undo the power of its teleological effect. . . . That would hold good also for the opposition between structures of horizontality and verticality that are at work in language. (p. 30)

Dislocated power is one discernable consequence when technology fragments linear narratives and other structures conventional to print documents (see "Retracing the Footsteps from Print to Digital" in this collection). Digital documents—as exemplified in The Wooster Group's CD-ROM—allow spectator-users to create narrative paths that are not "sequentially dependent" and thus can work against the "teleological effect," or the organizing structure that determines the set purpose for and outcome of narrative.

Though many have lauded the ways in which emerging forms of interactivity have substantially restructured our understanding of narratives and of the role of authors and readers, there are, of course, disadvantages to this restructuring. According to David Saltz (1997), interactivity must provide, in real time, a "sensing or input device" that can "translate certain aspects of a person's behavior into digital form." Thus the computer is made to generate output data according to a person's behavior and actions. The computer subsequently translates the data "back into real-world phenomena that people can perceive" (p. 118). Janet Murray (1997) identifies a key outcome of the responsiveness that digital media offer spectator-users: "When the things we do bring tangible results, we experience the second characteristic delight of electronic environments—the sense of agency" (p. 126). Consumers' desire for empowerment through interactivity—a symbol "of freedom and of progress" (Callens, 2002, p. 113)—relies on an uncritical utopian vision much like Faustus's desire for electric light.

The perplexities inherent in such a technologically utopian vision are exemplified by the spectator-user's relation with the CD-ROM performance document, as seen in Beloff's directorial choices. The entire composition of the CD-ROM focuses on the tension between notions of progress and agency. While computer technologies have been touted as bringing progress (including the ability to connect with people and places on a global scale), such progress is questioned when it begins to limit the amount of embodied interaction it affords its users. For example, at one point in the CD-ROM, there is binary code shining on one of the actors (signifying the rise of the computer), whose movements are stilted and restricted. This moment suggests that the computer—and by extension, the interactive CD-ROM—pushes humankind toward a *limited* type of freedom through a relation with technology. Interacting with The Wooster Group's CD-ROM involves the body to the extent that eyes scan the monitor, ears listen to the voice of Beloff, and fingers click on the mouse. Most of the physical body of the spectator-user is eliminated or marginalized in relation to its correspondence with the interactive document. The notion of freedom of movement is paradoxically inverted in the computer user, who moves very little as he or she clicks on the mouse to navigate virtually and connect globally. The CD-ROM, at every turn, raises questions concerning the amount of human agency we have in the face of technological determinism. We feel that we can connect across vast distances, yet we do not move anywhere. Thus, in one sense, the digital document, which seems to involve the audience more fully than the live event does, can give rise to an even more inert audience member.

Is "interactivity" nothing more than a masking of the authority of narrative sequence and authorial agency—that which places the spectator-user in a more or less passive mode of receiver instead of meaning maker? Marsha Kinder (2002) observes that although interactivity is at the core of many artistic forms (digital or otherwise), it remains illusory because it reinscribes the roles of author/reader:

> While all narratives are in some sense interactive in that their meanings always grow out of a collaboration between the idiosyncratic subjectivities of authors and audiences and the reading conventions of the respective cultures they inhabit and languages they speak, all interactivity is also an illusion because the rules established by the designers of the text necessarily limit the user's options. Interactivity thus tends to function as a normative term—either fetishized as the ultimate pleasure or demonized as a deceptive fiction. (p. 4)

Kinder's understanding of the binary (fetish or demon) implicit in the term *interactivity* is an insightful appraisal of the current cultural imagery surrounding interactivity and interactive media. While I do agree with Kinder that interactivity is illusory, I also think that such theories must take into account Murray's notion of the aesthetics of interactivity, which not only look to interactivity as a tool of agency for the spectator-user, but also offer the spectator-user "an aesthetic pleasure . . . to be experienced for its own sake" (1997, p. 128). Murray develops this idea further: "Because of the vague and pervasive use of the term *interactivity*, the pleasure of agency in electronic environments is often confused with the mere ability to move a joystick or click on a mouse. But activity alone is not agency. . . . Agency, then, goes beyond both participation and activity" (p. 128). Although interactivity is often confused with agency, interactivity might better be understood as involving an interface that offers spectator-users a *sense of* agency and thus an aesthetic pleasure. This variety of agency, the key to which is the possibility of interactivity, is framed by the constraints of an author, director, or code writer and is necessarily limited by the strictures of script and medium.

As seen in the CD-ROM, The Wooster Group and Beloff dramatize the tensions inherent in the term *interactive* and in the attempt to create an "interactive" performance. These tensions are performed on the digital stage of the CD-ROM. Though the interactive CD-ROM seeks to move beyond the authority of the text, it is in many ways still founded on the text of a play and at its core is controlled by a computer text limiting user choices and governing the results of those choices. Could the digital world of *Where Where There There Where* mirror the material world, which offers only limited choices mapped out by structural boundaries? As Marie-Laure Ryan (1991) writes, "On the map of narrative, as on the map of the world, boundaries are everywhere: Boundaries within the representing discourse, and boundaries with the represented system of reality; boundaries with gates to get across, and boundaries with only windows to look through" (p. 175). What happens to our understanding of progress when these structures and boundaries are exposed rather than veiled behind modes of production?

From the outset of the digital production of *Where Where There There Where*, precisely this question is staged as the spectator-user is posited onto the theatrical space of an abandoned factory. It is on this stage (with all its boundaries, from limited clickable spots to limited visual navigation) that The Wooster Group explores contradictions in our notions of progress and technology. Faustus sells his soul for electric light (represented in the CD-ROM by the factory, symbol of industrial progress). For Faustus, perhaps, there is no soul and there is no heaven to be damned from; there is no industrial or technological utopia that we can progress toward with our advancing technologies; instead, we are left with abandoned factories. Thus the tension between progress and obsolescence is staged in this document (which intentionally presents itself as a product of technological progress). Yet without fetishizing or demonizing interactivity, Beloff and The Wooster Group present an interface highlighting the discord between both approaches, situating spectator-users in a place where they are able to enter the scene of discord as a key character. Kinder notes that this entrance of the spectator-user onto the scene of the medium allows for an alternative view of interactive media: "One productive way of avoiding these two extremes [of fetishizing or demonizing interactivity] is to position the user or player as a 'performer' of the narrative" (2002, pp. 5–6). Put differently, if the aesthetic of The Wooster Group and Beloff constantly works to keep the spectator-user aware of the CD-ROM's status as a mediated document rather than as an immersive, interactive environment, in what way are spectator-users supposed to react to this constant reiteration of the medium? Through their inability to immerse themselves in this environment, spectator-users are meant to respond to the cultural expectations of interactivity and their inability to enact those expectations. The felt gap between expectation and actualization highlights the deeper authorial texts that script all our media, especially those forms in which authors seem to disappear, such as highly immersive and interactive media.

RESISTING IMMERSIVE INTERACTIVITY

The spectator-user whose interaction with the CD-ROM performance document actually alters the outcome of the performance is at once connected to the interactive document while distanced from it through the platform on which it is presented. That is, although the spectator-user is involved directly in creating the narrative structure of the document, he or she is always aware of the mediating platform of the CD-ROM and the QuickTime environment. The QuickTime movies and panoramas are nested into the user's computer screen and constantly highlight the framing nature of the performance. The CD-ROM constantly calls attention to the way it is produced, and each choice the user makes refers back to the mediating nature and binary logic of the computer system the performance is presented through. The Wooster Group's document resists "immersion" at every turn in a manner that evokes Brecht's Alienation Effect. This CD-ROM document does not allow the spectator-user to experience the piece as an "interfaceless interface"—a term used by Jay David Bolter and Richard Grusin (1999) in their work *Remediation: Understanding New Media*. Theorizing the "logic of immediacy," which defines immersion across media forms, Bolter and Grusin write, "The logic of immediacy dictates that the medium itself should disappear and leave us in the presence of the thing represented: Sitting in the race car or standing on a mountaintop" (p. 6):

> Our culture wants to both multiply its media and to erase all traces of mediation. . . . What designers often say they want is an "interfaceless" interface, in which there will be no recognizable electronic tools—no buttons, windows, scroll bars, or even icons as such. . . . Virtual reality, three-dimensional graphics, and graphical interface designs are all seeking to make . . . technology "transparent." (1999, pp. 5, 23)

Because the spectator-user of *Where Where There There Where* is never so immersed in the piece that he or she forgets that the computer is the tool mediating the digital document, there is little transparency here, little illusion of reality or totality in Faustus's digital world. High resolution is replaced with obvious pixilation and reduction in scale (as seen in the very small QuickTime movies that take up only a portion of the computer screen). The Wooster Group's choice of low-tech applications is intentional, designed to achieve an effect very different from that sought by Rudy McDaniel and Sherry Steward (see "Technical Communication Pedagogy and the Broadband Divide," in this collection) and by their colleagues in flight simulation and other high-tech industries. Vivian Sobchack (2003) notes that QuickTime's low-resolution aesthetic is rhetorically appropriate within a context that privileges, not immersive virtual reality, but the frame:

> Specific to the particular medium [QuickTime], such compartmentalization and grid work points not only to the larger order and framework of the surrounding "desktop," but also the smaller, hidden, and thus more "secretive" orders of the computer: Microchips, bits, and bytes. . . . Boxed fragments of photorealist images are compartmentalized further into smaller boxes yet—dissolving the personal meaning and contours of human memory and resolving them into the visible and controlled geometry that in-forms the underlying memory and structure of the computer itself. (p. 68)

The movies included on The Wooster Group CD-ROM, which seek to document an event and thus operate as a proxy for human memory, demonstrate how gaps in human memory cannot be filled satisfactorily through documentation. The nested character of the QuickTime movies makes it difficult for the spectator-user to watch the performance, for images are unclear and constant flicker and evident pixilation accompany each movie. In short, the CD-ROM document highlights its own inability to replicate the material world visually or to create a virtual reality—an inability to create a document that is just as "real" as the live event. Instead, the medium allows viewers to explore the tensions surrounding the notions of agency and interactivity in an age of progress (questions central to Stein's play), constantly framed.

To distance the spectator-user from immersion in the document, the interactive CD-ROM uses, in addition to low-resolution images, low-fidelity audio. Dialogue is difficult to hear and resembles the spoken word as heard on a very old, scratchy record. The piece also uses the image of the RCA dog that sits in front of the record player and listens to "His Master's Voice," a voice replicated, one is meant to infer, perfectly on the audio recording. In *Doctor Faustus Lights the Lights*,

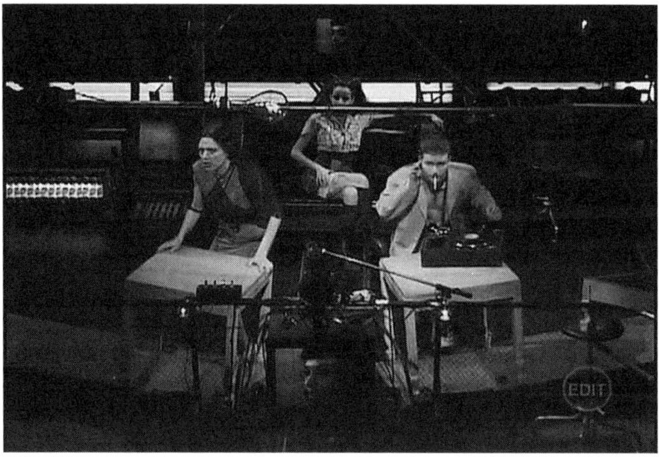

Figure 5. The *House/Lights* DVD. Pictured (l–r): Kate Valk, Suzzy Roche, and Ari Fliakos. The Wooster Group brings a Victrola phonograph on stage, as actor Fliakos plays the role of the dog listening to his master's voice. The interactive CD-ROM plays off this scene by creating a piece with low-fidelity audio mixed with low-resolution imagery, pointing toward the impossibility of bridging the gap between live event and digital documentation through high-definition or high-fidelity documents. (© The Wooster Group)

Stein stages the character of a dog that constantly says "thank you" and barks at the electric lights. This character translates into The Wooster Group's invocation of RCA Records' "His Master's Voice" dog. The image of the dog listening to a Victrola phonograph, recognizing the sound of the master's voice reproduced in high fidelity—used in the live performance of *House/Lights* as well as on the CD-ROM (Figure 5)—problematizes, as Callens (2002) notes, the "analogy between the dog's proverbial loyalty to his master and the phonograph's technical faithfulness to the human voice ('high fidelity')."

Callens continues this analysis:

> At first sight the HMV [His Master's Voice] dog's confusion of natural sound with its technological simulation, presumably because of the latter's alleged quality rather than the animal's (male-gendered) stupidity and servile devotion, simply signaled the confusion of Stein's dog between natural and artificial light, already conveyed in the stage production by its barking indiscriminately at video images of light bulbs and the moon. The transition between these images, though, was effected not by any illusion-enhancing montage but by a diaphragm-like cover manually swung from the top and bottom of two small monitors. Their resulting resemblance to magical boxes or high-tech devotional shrines, spelled overt deception more than realism, recalling once more the early days of cinema and the medium's primitive evocation of movement by a succession of stills thanks to the persistence of vision. (2002, pp. 124–125)

CONCLUSION

After the issuing of the interactive CD-ROM *Where Where There There Where*, which foregrounds the theme of access and comments constantly on its status as a document, a DVD document was produced of the staged version of *House/Lights* (presented on a digital platform that typically privileges high-resolution imagery). This three-disc DVD set, which includes the live performance on two discs and documentary footage of the tour of the performance on the third disc, presents the user with a much closer estimation of what the audience may have "actually" seen when attending the Performance Garage. For instance, viewers are allowed to change screen views, and thereby elements of the interactive nature of the CD-ROM performance are mirrored. The first view is a wide shot of the entire stage, which remains stationary. When the angle button on the remote control is pressed, the DVD switches to the second view—a mix of the various television screen displays. When *enter* is pressed, the viewer is taken to the third view—an edited version of the performance from various angles.

Still, we might question whether this DVD document, released 6 years after the interactive CD-ROM, is "closer" to documenting the performance of *House/Lights*. For instance, might it better represent the reality of the performance by simulating the spectator's ability to look around the various parts of the stage at will? Auslander argues that the notion of a "referent" in performance has become blurred. Documents and elements of performances (i.e., instant replay at football games or video monitors at concerts), initially designed to operate as "secondary elaborations of what was originally a live event, are now constitutive of the live event itself" (1999, p. 25). Although, as Peggy Phelan (1993) terms it, the "ontology of live performance" (p. 143) seems to reside in the live event itself, Auslander argues for a new understanding of the term *live performance* that is inextricably bound to the process of digitization. To assume that The Wooster Group's DVD operates as a secondary elaboration is to assume that there is a one-to-one, sequential relation between the live theatrical performance and the document conceived out of that performance. But the document is always embedded in the live performance, and the line between the live and the digitized is constantly blurred.

In the instance of the CD-ROM, the mode of production alters consumer expectations—the consumer does not merely buy a record of a performance but buys the platform on which the performance is produced. And in fact the user interface of the document can at times be just as important as the performance content of the document. This performance, presented on a platform meant to be viewed only on a Macintosh, is altered in the way it is received and "read" in an age—following Edison and Ford—of consumer loyalty as well as tensions in the face of progress and obsolescence. The QuickTime format points not only to the computer system, but much more specifically at the time of creation, to the Macintosh system. In fact, it can even be said that The Wooster Group's interactive CD-ROM performance was designed to be a performance by the Macintosh. The specificity of the platform, continually highlighted and framed by the performance, troubled the notion that such a document might be eternal or even long lasting in an age when computer platforms and files become obsolete as the engine of progress produces emerging machines and new formats. The Wooster Group's CD-ROM *Where Where There There Where* is a performance experienced not only *in* but also *with* a very specific platform in a very specific moment in technological history. Continually foregrounded is the role that these media play in our interactions with documents. Users engage the digital with the keen understanding that they, fully aware of the mediating interface and structure's specificity, are always distanced and removed from any true sense of control over the document.

The relation between the two performance documents of The Wooster Group becomes a performance of its own—challenging the idea that the document operates as a proxy for the fleeting live performance and attempting to anchor it in some way to a repeatable and consistent format. Instead, as seen in the relation between *House/Lights* and *Where Where There There Where*, the "live" performance does not take primacy over the document, but instead the document creates a new theatrical space. This is true not only with The Wooster Group's performance of the document through use of digital media, but with all forms of documentation. Documents always create space into which new interpretations will be applied. That is, documents always constitute a new performance. The Wooster Group, playing off this understanding, creates a CD-ROM that operates as a performance of the document, rather than as a document of the performance. Beloff goes so far as to say that *Where Where There There Where* is not to be understood as a document of the play, but as a satellite work. Despite differences between the "live" performance and its documentation on CD-ROM and DVD, any transparency between "live" event and digital production is an illusion. The Wooster Group exploits this inherent slippage, creating a document that troubles the notion of documentation, a theatrical product that shifts noisily into the realm of commodities available through the media of CD-ROM and DVD. For The Wooster Group, the line between live and digitized must be understood as unstable, and authority as perpetually under construction.

REFERENCES

Adato, P. (Producer and Director). (1970). *Gertrude Stein: When this you see, remember me.* Carlsbad, CA: CRM Films.

Auslander, P. (1999). *Liveness: Performance in a mediatized culture*. New York: Routledge Press.

Beloff, Z. (n.d.). *Doctor Faustus and the universal machine*. Retrieved October 14, 2006, from http://www.zoebeloff.com/where/ideas.html

Boal, A. (1985). *Theatre of the oppressed*. New York: Theatre Communications Group.

Bolter, D., & Grusin, R. (1999). *Remediation: Understanding new media*. Cambridge, MA: The MIT Press.

Brecht, B. (1957). *Brecht on theatre: The development of an aesthetic*. (J. Willett, Ed. and Trans.). New York: Hill and Wang.

Callens, J. (2002). Going public, performing Stein. In B. Ozieblo & M. López-Rodríguez (Eds.), *Staging a cultural paradigm: The political and the personal in American theatre and drama* (pp. 113–130). Brussels: P.I.E.-Peter Lang.

Case, S. (1996). *The domain-matrix: Performing lesbian at the end of print culture.* Bloomington: Indiana University Press.

Gendrich, C. (1998). Performance review: *House/Lights. Theatre Journal, 50*(3), 381–382.

Haye, B. (1984). The Wooster Group: Traditional plays in a radical experimental context. In O. Zuber-Skerritt (Ed.), *Page to stage: Theatre as translation* (pp. 165–182). Amsterdam: Rodopi.

Kilpatrick, D. (2000). Catching up with Gertrude Stein's *Faustus* and the Wooster Group. In E.Voigts-Virchow (Ed.), *Mediatized drama, dramatized media* (pp. 67–76). Frankfurt: Wissenschaftlicher Verlag Trier.

Kinder, M. (2002). Hot spots, avatars, and narrative fields forever: Buñuel's legacy for new digital media and interactive database narrative. *Film Quarterly, 55*(4), 2–15.

Murray, J. (1997). *Hamlet on the holodeck: The future of narrative in cyberspace.* Cambridge, MA: The MIT Press.

Phelan, P. (1993). *Unmarked: The politics of performance*. New York: Routledge Press.

Ryan, M. (1991). *Possible worlds, artificial intelligence, and narrative theory.* Bloomington: Indiana University Press.

Saltz, D. (1997). The art of interaction: Interactivity, performativity and computers. *Journal of Aesthetics and Art Criticism, 55*(2), 117–127.

Sobchack, V. (2003). Nostalgia for a digital object: Regrets on the quickening of QuickTime. In J. Shaw & P. Weibel (Eds.), *Future cinema* (pp. 66–73). Cambridge, MA: MIT Press.

http://dx.doi.org/10.2190/CWDC5

CHAPTER 5

Digitizable Cultural Capital: Anticipations of Profit in the Web Market

John B. Killoran

OVERVIEW

This chapter explores the relatively small population of individuals who self-publish résumés on the Web. Drawing on a survey of 100 authors of such résumés, I show that these authors tend to work in a narrow range of white-collar occupations, and especially in occupations that involve producing content that can be displayed readily and appealingly in digital form. Compared with authors who post less, authors who post more and who are more likely to integrate content with their Web résumés perceive their résumés as more useful to them. To explain these findings, I use Pierre Bourdieu's work on symbolic profit and cultural capital to argue that publishing in a digital medium is rational for those who can publish more digitizable cultural capital. Turning to pedagogical implications, I discuss how professional communication students might develop their digitizable cultural capital by publishing portfolios to accompany their Web résumés.

Like much digital usage research that has focused on the activities of consumers, a periodic study conducted by the U.S. National Telecommunications and Information Administration (2004) explores activities such as consumer searches for and purchases of Internet products and services. Among scholars of composition, literacy, professional writing, and communication, however, *digital literacy*, including literate Internet usage, is understood to include not only consuming but also producing a broad range of digital discourse (e.g., Grabill, 2003; Selfe, 1999). As Jan Van Dijk (2005) observes, "The difference between a relatively active or creative and a relatively passive and consuming type of usage of the Internet is a much-neglected aspect of the digital divide" (p. 115). In a meta-analysis of research about gender and the Internet, for example, Cindy Royal (2005) finds few studies examining women as producers of Internet content, a role she characterizes as potentially empowering.

What is more, the type of online content being produced may be narrow. Amanda Lenhart, John Horrigan, and Deborah Fallows (2004) report that although 44% of U.S. adult Internet users have contributed ideas or files of their own to the Web or to newsgroups, these Internet users participate in fewer than two of the dozen types of potential online content-creation activities (p. 2). That creating and maintaining a Web site of one's own is not among the most widespread online content-creation activities suggests that, despite the millions of Web sites in existence, site production remains a rather exclusive pursuit.

No doubt differences in Web authoring skills are a key factor explaining why certain individuals produce Web sites of their own while most others do not. But even as the cost of computer technology and online access declines and the usability of Web design software increases, only a minority of the Web-consuming population takes advantage of the opportunity to publish personal sites. Thus Web authoring skills and economic resources alone may be necessary but not sufficient prerequisites for self-publishing. Understanding what is sufficient could help professional communication students, already Web consumers, to assess whether it is worth their time and labor to become Web producers. I propose in this chapter that the prospect of Web publishing invites primarily those whose investment is likely to be symbolically profitable in the Web market.

Focusing on a specific form of digital divergence characterizing the Web publishing population, this chapter investigates the commonplace professional genre of the résumé and explores which occupational groups self-publish résumés on the Web. At a time when the Web is being hailed as a publishing boon for ordinary people, the realities of self-publishing Web résumés suggest uneven opportunities for would-be publishers; this fact raises the question of just how viable Web production really is in the universal sense often held out as a goal in discussions of digital literacy. Whose education, employment, skills, and other achievements are more or less readily digitizable, and hence who is more or less able to invest their résumé, and by extension other discourse, profitably in the Web market?

Résumés have, of course, been produced widely in print form by individuals across most occupations. If there were no digital divergence in regard to the self-publishing of Web résumés, we would expect that a similarly diverse range of résumés would be available online. Yet this study's analysis suggests that most authors of Web résumés may belong to a narrow range of occupational clusters. The striking disparity in the use of Web résumés across occupations should compel the attention of researchers and instructors of academic fields that are defined in part by the occupations they serve (e.g., technical communication).

To explain this phenomenon, I draw on the theoretical perspectives of social and linguistic theorist Pierre Bourdieu (1986). Résumés summarize what Bourdieu would call our *cultural capital*, or our noneconomic wealth in forms such as knowledge, skills, and educational degrees. One way we can profit from cultural capital is by offering it in exchange for employment and thereby a paycheck (or economic capital); thus we are advised to list our cultural capital on our print résumés. Whereas several researchers have explored print résumés, as well as what Bourdieu would characterize as the cultural capital of these documents

(Culwell-Block & Sellers, 1994; Harcourt & Krizan, 1989; Hutchinson & Brefka, 1997; McDowell, 1987; Moody, Stewart, & Bolt-Lee, 2002; Popken, 1999), few researchers have focused on digital résumés or, specifically, on résumés self-published on the Web. And although certain professional communication textbooks caution students against relying on a self-published Web résumé to do the hard work of seeking employment (e.g., Anderson, 2003, p. 38; Oliu, Brusaw, & Alred, 2004, p. 604), many individuals make the effort to create, post, and maintain a Web résumé, presumably with some anticipation of symbolic, if not financial, profit for their efforts.

The data discussed in this chapter suggest that Web résumé publishers who rate their Web résumés as more useful tend to come from a cluster of occupations whose work involves producing cultural capital that can be displayed readily and appealingly on an online digital interface, by means of text, graphics, sound clips, video clips, and computer scripts—content that I refer to as *digitizable cultural capital*. I infer two other key findings from the data: Those who rate their Web résumés as more useful tend to have posted more digitizable cultural capital and to have integrated it with their résumés. In short, the digital divergence observed in this study may be understood in terms of the likelihood that a career path will promote production of such capital. Because of the increasingly digital nature of professional work, aspiring professional communicators might ask themselves whether they have or can readily acquire the discourse needed to participate profitably in the Web market. This study offers cause for both optimism and concern.

To explore the role of digitizable cultural capital, I first review Bourdieu's quasi-economic perspective on the production of discourse, and specifically his notion of *cultural capital*. I then explain how such capital, if it is to be communicated on the Web, must be readily accessible in digital form. To illustrate the influence of digitizable cultural capital on Web publishing, I draw on a survey of 100 authors of self-published Web résumés, as well as on an analysis of their résumés and adjoining Web pages. This analysis shows that those who have published their résumés on the Web tend to be in occupations involving the production of digitizable cultural capital. I suggest that among Web consumers it may be that those with more digitizable cultural capital are more likely to perceive self-publishing as symbolically profitable and are more likely to become Web producers. In addition, those who publish more of such capital and who integrate it into their Web résumés tend to perceive their résumés as more useful. Finally, turning to pedagogical implications, I show how professional communication students, individuals likely to benefit economically from these activities, can respond to the challenges of acquiring and can further develop digitizable cultural capital by publishing their résumés and portfolios on the Web.

BOURDIEU ON SYMBOLIC PROFIT AND CULTURAL CAPITAL

According to Bourdieu's economic reasoning, the discourse we produce is assessed not so much as a source of meaning but as a sign of wealth in discursive markets (1991, p. 66). We in turn orient our discourse to the opportunities available, investing where we anticipate we might gain symbolic profit and otherwise refraining from investing (1977). *Symbolic profit*—a concept that for Bourdieu (1986) extends well beyond economic profit, to include accruals of recognition, acknowledgment, status, prestige, honor—can be produced through investments of various forms of capital:

- *economic capital*, or our financial wealth;
- *social capital*, or the social network from which we draw obligations; and
- *cultural capital*, or our store of knowledge, skills, and other related acquisitions.

Bourdieu subdivides cultural capital into three states:

- *embodied*, which grows and declines along with the body and mind, such as one's physique, deportment, and knowledge and skills;
- *objectified*, which is constituted by one's physical possessions, such as housing and clothing, art, information and communication technologies; and
- *institutionalized*, which is sanctioned by organizations such as degree-conferring academic institutions.

Such an expanded perspective on capital can offer a useful corrective to explanations of "the digital divide," some of which localize complex social issues as solely or mainly economic, as the editors of this anthology point out in their introduction.

As Bourdieu explains, we are not endowed with meaningful stores of cultural capital at birth; rather, we develop them primarily through family upbringing and socialization (1986). Families already well endowed with cultural capital can nurture such capital in their children, for instance, by reading to them in the home. In contrast, children of less well endowed families may have to wait until they begin school before learning to read. Bourdieu emphasizes that most cultural capital requires considerable time to develop, as with the increasing levels of literacy needed to read, to appreciate, and eventually to write literature; well-endowed families are comparatively able to provide children with the free time required to develop such cultural capital. Alongside demographic factors such as class, race, gender, and age—factors relevant to the issue of digital divergence—cultural capital shares some of the permanence that makes inequitable digital usage such a recalcitrant social challenge. An understanding of divergences in the publishing of Web-based résumés, one that both empirically documents and theoretically understands divergences in terms of cultural capital's role, may advance scholarly and instructional discussion of occupation. That is, the mutually supportive combination of data and theory may suggest workable solutions that a demographic approach cannot.

Such solutions would start with an understanding of how and why we invest our capital on the Web, and further may explain how Web production gives rise to particular divergences largely absent in print forms. According to Bourdieu, our portfolio of various forms of capital determines which markets we can invest in profitably (1991). Although much of our economic and social capital cannot be displayed readily in the Web market, cultural capital often can. To be communicated on the Web, each kind of cultural capital must be representable in a digital form.

- *Embodied.* Whereas one's body and mind are not themselves digital, the body can be represented digitally through photographs, audio files, and video clips; and the mind can express its knowledge and skills digitally through writing, designing, and publishing a Web site.
- *Objectified.* The Web site itself is part of one's objectified cultural capital. Apart from such digital possessions, most physical objects cannot themselves be transmitted on the Web, but as with embodied cultural capital, they can be represented through photographs, audio files, and video clips.
- *Institutionalized.* The capital that organizations confer has traditionally been represented through official terminology (e.g., PhD), quantitative measures (e.g., grades), and publicly recognized symbols (e.g., graduation regalia), and such capital can be represented readily on the Web through words, numbers, and pictures.

In Bourdieu's terms, the résumé is a primary discourse by which people exchange such embodied cultural capital as their knowledge and skills, and such institutionalized cultural capital as their educational degrees, for the economic capital arising from employment, thereby profiting financially and symbolically from their invested capital. If, as Bourdieu maintains, we invest discursively in those markets in which we anticipate that our discourse will gain us symbolic profit, then individuals who have digitizable cultural capital, especially given the competition to secure employment and to develop careers in capitalist economies, may be especially likely to invest in the Web market—that is, to publish a Web site exploiting their cultural capital. After all, many discourses and genres have been migrating from print to the comparatively populist Web, and a Web résumé has substantial potential to supplement the terse one-page print résumé and to reach new audiences. If, however, a limited number of print résumé producers are becoming Web résumé producers, then a difference may have emerged on the Web where none seems to exist in print—making the Web résumé an opportune genre through which to study divergent Web production.

This chapter explores how differences in cultural capital might explain who can and who cannot invest profitably in the Web market. To illustrate the influence of digitizable cultural capital in Web publishing, I draw here on a study of individuals who have self-published their Web résumés. To understand these Web résumé producers and the role of cultural capital in their Web investments, I both surveyed individuals who had self-published their Web résumés and examined their résumés and sites.

METHODS

Collecting a random or representative sample of self-published Web résumés is not feasible because many such résumés remain hidden from search engines; consequently, I aimed instead to collect a diverse, nonprobability sample of Web résumés. To collect the majority of my sample, I used the AltaVista search engine, which, unlike search engines such as Google, does not use a ranking formula giving precedence to popular sites. Additionally, AltaVista permits country-specific searches, a feature that enabled me to target countries including Brazil, Canada, India, Ireland, and the United Kingdom, and thereby avoid a sample dominated by U.S. participants and sites. To expand my sample further, I conducted searches within the more proletarian population of sites posted for free by Geocities and Tripod. Collected in the spring of 2003, the resulting sample, entirely in English, is, of course, not representative of the elusive Web résumé population of that time, but its diversity can unearth issues that merit further exploration.

Prospective participants were e-mailed a solicitation to participate in the survey, together with the survey itself. The survey asked, among other things, that respondents rate how useful their Web résumés had been to them. After I received the completed surveys, I downloaded the respondents' Web sites to provide a stable corpus for further analysis. Three sites were dropped at this stage of the study because the site or its server no longer was accessible. Thus, when discussing the Web résumés themselves, I draw on a sample of 97 Web sites, though when discussing the survey results alone I draw on the full survey sample of 100.

Three coders independently analyzed the 97 Web résumés in the sample to determine their authors' occupations. Because almost all résumés appeared within the context of a larger Web site, the coders also examined the site files linked to each résumé to determine the best occupational fit. The classification scheme was adapted from the Standard Occupational Classification (SOC) system devised by the Bureau of Labor Statistics (2004), a unit within the U.S. Department of Labor (DOL). The SOC consists of a hierarchy of occupation types in which the top level comprises 23 major occupational categories (see Appendix A for a list). Whereas the DOL characterizes the SOC as "reflecting the current occupational structure in the United States" (Overview section), which is arguably the world's most developed economy, the international population that publishes résumés on the Web would be expected to have at least the same qualifications: Access to a computer, the literacy to write a résumé, and the computer literacy to design and to post it on the Web. The SOC therefore offers a reasonable classification scheme for this international sample. Moreover, by relying on the 23 major categories and not on the more numerous subcategories, the classification scheme is sufficiently universal to accommodate local differences in occupational practices. Roughly half of the 23 occupational categories could be characterized informally as white collar and the other half as blue collar.

In classifying occupations, the coders followed the DOL's guidelines: "Occupations are classified based upon work performed, skills, education, training, and credentials. . . . When workers may be classified in more than one occupation, they should be classified in the occupation that requires the highest level of skill" (2004, Classification Principles section, point #7). Applying these guidelines to the classification of Web résumés presented challenges, however, for the SOC is limited to "occupations in which work is performed for pay or profit" (point #1), a definition excluding, among other things, the unpaid educational labor of students.

The coders attempted to consider career trajectory broadly, rather than focusing on a respondent's current occupation. For instance, the sample included several full-time students who were studying computer science, were clearly skilled with computers, and were doing part-time or freelance work as computer programmers or Web designers. In such instances, the coders judged the career trajectory to be computer related and classified the occupation accordingly. Attempting to classify all students by their education or part-time work was not always feasible, as their field of study might not lead to a specific career track, or their part-time work might not represent their career potential fairly. Hence, the classification "student" was added to the SOC's 23 major occupational categories, making a total of 24 categories, to accommodate résumé writers without a clear career track whose most skilled occupation was full-time study.

Each résumé was assigned, based on the coders' classifications, one primary occupational category. When the three coders' classifications differed, the occupational category was determined by the majority classification. Coders disagreed at times, especially in regard to the sample's many computer-related workers whose occupations could be categorized as, obviously, a computer occupation, but one also connected to design (e.g., Web designer), engineering (e.g., computer hardware engineer or electrical engineer), management (e.g., systems manager), or installation, maintenance, and repair (e.g., computer equipment installer). Because of the novelty and dynamism of the computer field, several respondents had recent experience in two or more areas. In one instance, all three coders classified a résumé within a different occupational category, so that résumé is not included among the primary occupation results.

DIGITIZABLE CULTURAL CAPITAL

Although this chapter focuses on the sample's occupational distribution, I should mention that the sample also may illustrate gender differences in production of the résumé genre. As Royal (2005) observes, since U.S. women have reached parity with men in terms of access to the Internet, recent Internet research about gender has turned to explore more specific inequities, such as inequitable patterns of production. Indeed, the sample I obtained was overwhelmingly male (81%), a striking disparity echoing the demographics of personal homepage authorship from the later 1990s (Dominick, 1999; Killoran, 1999). By contrast, Lenhart and colleagues (2004) found that U.S. adult creators of online content of *any* kind are nearly balanced by gender, although Lenhart and Mary Madden (2005) found that the gender balance among U.S. teen creators of online content shifts across types and genres of content. Similarly, Susan Herring, Inna Kouper, Lois Ann Scheidt, and Elijah Wright (2004) found that authorship of the blog genre is fairly balanced by gender overall but not across blog subgenres. It could be that the market offers different opportunities for symbolic profit depending on the genre and the nature and extent of one's capital, which in turn could depend to an extent on gender. Other factors, including the sample's international makeup and occupational distribution, may have contributed to the gender imbalance.

Similarly, the sample's age distribution shows a bulge among those in their twenties and thirties (see Appendix B). This distribution roughly corresponds with that of adult U.S. Web-content creators (Lenhart et al., 2004) though it almost entirely bypasses the majority of online U.S. (and presumably international) teens who also are Web content creators (Lenhart & Madden, 2005). Teens, of course, often do not yet have the need to create a résumé or the cultural capital to warrant one.

The occupational distribution of those who self-publish Web résumés suggests that the extent to which individuals perceive the Web market as potentially profitable—symbolically and financially—may be based on the type of cultural capital they possess. In fact, two-thirds ($n = 66$) of the résumés studied were accommodated by just two occupational categories (see Appendix A):

- computer and mathematical occupations ($n = 39$, all in computer-related occupations, not mathematics), and
- arts, design, entertainment, sports, and media occupations ($n = 27$, roughly half in such computer-intensive occupations as graphic design or Web design).

Three additional categories accommodated most of the rest of the résumés:

- management occupations ($n = 9$),
- architecture and engineering occupations ($n = 9$, almost all engineering), and
- education, training, and library occupations ($n = 5$, mostly academic).

The sample included 20 students although only one was categorized within the *student* occupational category because the majority were studying and working part-time or full-time on well-defined career tracks.

Surprisingly, certain white-collar professionals who usually have access to computers in their workplaces and who generally are perceived to command considerable cultural and economic capital, such as professionals in the legal and healthcare fields, were almost entirely absent from the sample, and no participant was employed primarily in blue-collar occupations although a couple of participants performed blue-collar in addition to their primary work. For instance, one computer technician and Web manager also worked as a beekeeper. Another participant, who worked primarily in the white-collar computer occupations category, evidently also worked in computer installation and repair, which was included within the blue-collar category of installation, maintenance, and repair occupations. Lending further evidence to the absence of blue-collar occupations, only 25 Web résumé authors did not already have the equivalent of a four-year college degree, and many of those were college students on track to obtain such a degree. Twenty-nine had completed a graduate degree, including eight PhDs, and one MD who had become a professor. The sample thus seems noticeably more educated than, say, the general population of adult U.S. online content creators (Lenhart et al., 2004), half of whom do not have college degrees.

Because the sampling method was not designed to gather a representative sample of the world's self-published Web résumés, the precise occupational or educational distribution of the sample should not be generalized to the Web as a whole. It seems unlikely, however, that the resources used in sampling—primarily AltaVista, and secondarily Geocities and Tripod—would selectively favor the résumés of two occupational groups among the endless kinds of Web pages they search or host. The strikingly limited occupational distribution of the sample consequently offers suggestive evidence that the population of Web résumé publishers may be markedly different from the population of the world's workers, and indeed different from the population of the world's relatively affluent English-speaking workers.

An initial glance at the sample's two dominant occupational groups might suggest that skills with computers are the sole determining factor accounting for who does or does not post résumés to the Web. For instance, the sample's most populated occupational category, computer and mathematical occupations, was made up almost exclusively of software programmers, systems administrators, and other computer-intensive workers. Similarly, the sample's second-most populated occupational category—arts, design, entertainment, sports, and media occupations—included many computer-intensive workers such as graphic designers and Web designers.

A deeper examination, however, suggests that more than computer skills are at play here, as seen with the second-most populated occupational category. This category included not only computer-intensive workers, but also twelve participants in occupations not traditionally perceived as computer intensive: three actors, two writers, two painters, and others in the performing and fine arts. An additional three participants in this occupational category balanced computer-intensive design work such as Web design with less computer-intensive artistic work such as printmaking and music. In total, of the 27 participants in this occupational category, more than half (15) worked wholly or partly in occupations not traditionally regarded as computer intensive.

To assess the efficacy of Web résumés, one survey question asked respondents to rate the overall usefulness of their Web résumés, on a scale from 0 (not at all useful) to 4 (very useful). The average rating of the sample was 2.56. But even though the collective computer involvement of the arts, design, entertainment, sports, and media group was diluted by the substantial presence of less computer-intensive artistic and cultural workers, participants in this occupational category reported that their Web résumés were significantly more useful to them than participants in the computer and mathematical occupations category did (2.96 vs. 2.31, respectively; $p < 0.05$).

This curiously strong showing among an occupational group that has only a mixed involvement with computer use, together with the absence of certain white-collar occupations involving regular computer use, suggests that factors other than computer skills may play a key role in determining who self-publishes Web résumés. One characteristic shared among those in the occupational category of arts, design, entertainment, sports, and media, whether the computer-intensive graphic and Web designers or the less computer-intensive artistic and cultural workers, is that their cultural capital is readily and appealingly displayable on the Web. For the computer-intensive graphic and Web designers, such digitizable cultural capital includes images of their design work as well as the design of their own Web sites. To illustrate, Figure 1 reproduces the elegantly designed homepage of graphic artist and designer Athena Cooper, which introduces her portfolio and features a thumbnail screen capture of one of her recent Web design projects. Her site includes more than three dozen images of her work across such diverse areas as Web and graphic design, computer modeling and texturing, and drawing and painting, as well as more than a dozen video clips showcasing her animation and video editing work.

For less computer-intensive artistic and cultural workers, digitizable cultural capital might include video clips of acting, photographs of art works, excerpts of writing, and so forth. Indeed, the sample includes three actors. One, a part-time actor who also works as a technical writer and trainer, posts two different résumés: His acting résumé includes a shoulder-length photograph accompanied by such statistics as his height and weight whereas his writing and training résumé appears without

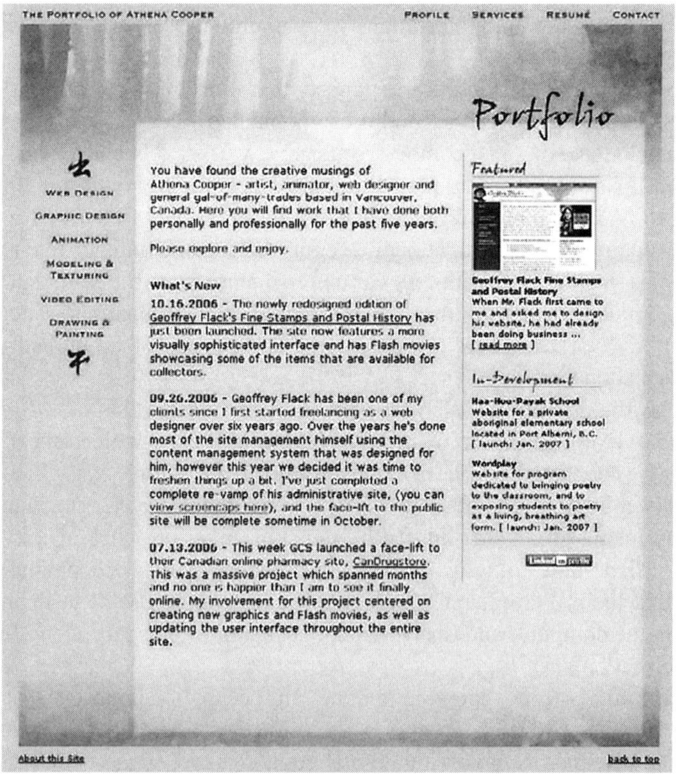

Figure 1. The homepage of graphic artist and designer Athena Cooper, at http://athena.adwin.ca, introduces her portfolio and features a thumbnail screen capture of a recent Web design project. Used with permission.

the photograph but links to a writing sample outlining his conception of effective communication. A second actor includes seven photographs of himself on his résumé page alone, as well as links to other pages featuring another 27 photographs from his theater and film performances. A third actor includes links from his résumé to a page of photographs of himself in one of the movie productions in which he played a lead role, as well as to film sites for three other movies he participated in. Additionally, his homepage links to pages devoted to photographs of him performing in various productions, for a total of 41 photographs of his acting work across the site. In response to a survey question, this actor explained how he believed visitors would be viewing his site:

> Basically, someone will either stumble onto or be directed via search engine to my Web site. When they want to look more closely at my qualifications, my résumé provides that. It appears that about 25% of all folks who have visited my Web site look at my résumé. About 50% look at my headshots and character shots on my pictures page. . . . About 20% read my News page, which is updated weekly.

In response to another question, he explained that posting digitized cultural capital is now standard for actors: "I see [my Web résumé] as an absolute necessity in our changing world. Things happen too fast for the mail. . . . Many of the casting agencies are now using [e-mail] exclusively." Similar practices are becoming common among other arts and cultural workers as well. Madden (2004), for instance, found that a large majority of the U.S. musicians she surveyed had posted their own music online.

In contrast, many white-collar workers whose occupations involve regular computer use and who are perceived to be endowed with considerable cultural capital in employment markets outside of the Web tend not to have the kind of cultural capital that is readily and appealingly displayable there. We might wonder, for instance, what material doctors or lawyers or accountants might post on a Web site to accompany a résumé. Included in the sample I studied was only one health-care practitioner: A dermatologist whose site included 36 files of various types (text, graphics, scripts), but most directly related not to his practice as a dermatologist but to his work as an assistant professor with a bioinformatics background, which involved academic research and a computer application he had developed for dermatologists. The sample's only participant in a community and social services occupation was a graduate student of applied clinical counseling whose site included 75 files, but these consisted of hobbies and interests, personal news, inspirational quotes, and photos of family and friends—a personal homepage, with little material related directly to her desired occupation. The sample's only participant in a business and financial services occupation was a purchasing manager whose site consisted of two files: Her résumé in different file formats. Apart from their résumés, these educated white-collar workers posted little that represented the cultural capital endowed by their occupations, perhaps because such cultural capital was not readily or appealingly digitizable.

Such differences among occupations suggest that the Web résumé, to be worthwhile, may need to be accompanied by a wide range of genres, all digitized. This finding may call into question conventional assumptions about the amount of content that ideally will reach potential employers by means of a print résumé accompanied by only a cover letter. At the very least, teachers and students might explore the possibility that, depending on their profession, job seekers should consider the possibility that a genre-rich Web site to which readers of a cover letter/résumé might be directed has the potential to enhance their employment opportunities. When asked to distinguish between their print and Web résumés, many respondents mentioned a variety of additional genres, such as the text of their publications, screen shots of their Web design work, thumbnail images and photographs of other visual work, sound clips of their music, or downloads of their computer scripts. Typical was the comment of one respondent who explained that her Web résumé was only one feature of her site, which she characterized as "more portfolio than strictly a résumé." Another respondent, who completed a graduate degree specializing in educational technology and worked in a college technology center, explained, "This résumé is a small part of a significant amount of material that I have posted to the Web. It is in no way an emphasis of my online work." His online work comprised 469 files, many devoted to his academic portfolio, including hypertext essays and material related to his field, such as a communications primer for fellow college instructors. Some respondents explained that their sites' purpose was to showcase their work. For instance, a Web developer and Web site manager whose technologically sophisticated site even included a page-long colophon explained, "I see [my résumé's] future use entirely in conjunction with the rest of my site. . . . I would use the whole site to demonstrate my abilities/skills to potential employers or clients." Another respondent mentioned that her Web résumé included

> photos, sound clips, links to other sites (thumbnails), overall design and color that a paper copy doesn't have, and generally more detail on each subject. *Everything* is included and organized, whereas when I hand out a one or two page résumé to an employer, I cull it to reflect what I think that employer is looking for rather than give them the whole thing and let them wade through it. I use the online résumé to grab the content for other more specific résumés that I may hand out.

Some participants also suggested that it was the impracticality of showing such digitizable cultural capital with a print résumé that had led them to create a Web site. For instance, an aspiring photographer pointed out, "The web gives me the opportunity to put out my portfolio, which I would not be able to do otherwise." Another respondent, a software developer and Web designer, explained, "When designing my résumé, I wanted to convey a lot more information than is on the standard print résumé of one or two pages. For example, I provide a dozen samples of source code I have written, in their entirety—something that would have required several dozen pages [in print]." Having digitizable cultural capital could thus be one criterion in Web résumé publishing, encouraging primarily those with such capital to invest their time and labor in a Web site, and perhaps discouraging those with too little to make the Web investment profitable.

If, as Bourdieu maintains, discursive markets award symbolic profit to those who invest their cultural capital, then through analyzing the data on Web résumés, we might be able to measure the perceived profitability of digitized cultural capital invested in the Web market. Thus I sought to correlate quantitative indicators of the cultural capital on participants' sites with participants' ratings of their Web résumé's usefulness (a Web résumé's usefulness rating being a kind of measure of perceived profitability, whether symbolic or financial, in the Web market). For quantitative indicators of digitized cultural capital, I tabulated the following three features:

- The total number of site files, which could be an indicator of the author's total stock of digitized cultural capital.
- The number of unique links from within the Web résumé to other files on the site, not including links at the periphery of the résumé in the site navigation bars. These links could be an indicator of the author's résumé-related digitized cultural capital.
- The number of unique graphics within a Web résumé. These graphics could be an indicator of the résumé-related digitized cultural capital depicted within the graphics, such as artwork depicted within photographs.

All three of these variables could also be indicators of an additional kind of résumé-related digitizable cultural capital: The skills to create Web files, links, and graphics. Because the resulting set of numbers for each feature included obvious outliers that could skew the representation of a typical Web résumé, such as the few sites that numbered over a thousand files each, I converted each feature's set of numbers into a high-to-low ranking from 1 to 97 (for the 97 sites in the sample). The entire sample was then divided into two roughly equal subgroups: A "less-useful" subgroup of those who rated their Web résumé less useful than average (a level of 0, 1, or 2), and a "more-useful" subgroup of those who rated their Web résumé more useful than average (a level of 3 or 4). Thus, between-group comparisons were made of rankings of files, links, and graphics to determine whether these indicators of digitized cultural capital were associated with the résumé's perceived profitability in the Web market.

For each of the three quantitative variables, the Web sites of the more-useful subgroup showed higher rankings of digitized cultural capital than did those of the less-useful subgroup (see Appendix C):

- The median number of site files in the more-useful subgroup was more than double the median number in the less-useful subgroup (178 and 75, respectively), and the difference between the two subgroups' rankings was statistically significant ($p < 0.05$). Hence, authors' ratings of their résumé's usefulness were related to their total stock of digitized cultural capital.

- Approximately nine out of every ten sites included résumés with links to other files within the site, an indication that most authors were not simply dumping their print résumés online but were taking advantage of the Web's affordances to integrate their résumés with other relevant files. Though some of these links were primarily functional, such as links to contact information, others linked to files that clearly represented digitized cultural capital. The median number of such links within the résumés of the more-useful subgroup was slightly greater than that of the less-useful subgroup (8 and 6, respectively) though the difference between the two subgroups' rankings was not statistically significant ($p > 0.05$). Of course, the site navigation bars at the periphery of the résumés offered alternative ways to link to other site files, so the limited numbers of links within the résumés themselves should not be taken to indicate that résumés are otherwise isolated from their sites' other digitized cultural capital.
- Approximately five out of every six sites included résumés with graphics, indicating, as with links, that most authors were attempting to take advantage of the Web's affordances, at the very least by displaying digital design skills, a kind of digitizable cultural capital. Though some of these graphics served primarily to enhance page design, others clearly depicted résumé-related digitized cultural capital, such as photographs depicting an author's visual work. The median number of unique graphics within the résumés of the more-useful subgroup was more than twice the median in the less-useful subgroup (5 and 2, respectively), and the difference between the two subgroups' rankings was statistically very significant ($p < 0.01$). Hence, authors' ratings of their résumé's usefulness were related to the digitized cultural capital depicted within résumé-related graphics.

Overall, these results suggest some association between the quantity of digitized cultural capital and a résumé's perceived profitability in the Web market and are thereby broadly in line with what Bourdieu's theory of cultural capital and symbolic profitability would predict. To be clear, such results do not indicate a direct causal connection between digitized cultural capital and a résumé's perceived profitability and should not be taken to suggest that simply adding more stuff to a résumé will increase its profitability. Quantity likely cannot substitute for quality. Other factors may well play a more significant role in these results. Indeed, both the quantities of digitized cultural capital that authors publish and their Web résumés' perceived profitability might be accounted for instead by authors' occupations. For instance, in comparing the sample's two largest occupational groups, we saw that participants in the arts, design, entertainment, sports, and media occupational category rated their Web résumés significantly more useful than did participants in the computer and mathematical occupational category. If we now compare these two occupational groups according to quantities of the three Web résumé features, we find trends very similar to those found when comparing the more-useful and less-useful subgroups. Compared with the résumés of those in the computer and mathematical occupational category, the résumés of those in the arts, design, entertainment, sports, and media occupational category ranked higher in quantities of all three features, though the difference was statistically significant only for the number of unique graphics within their résumés (see Appendix D). Hence, it is possible that the Web market is perceived as awarding greater profitability not so much for posting more content per se but for holding an occupation whose content is more likely to include digitizable cultural capital. Indeed, comparisons of quantities of the three features across the more-useful and less-useful subgroups within the *same* occupation failed to identify any significant differences. That is, participants with computer occupations, for instance, seemed not to perceive that they profited more by posting more content whereas their perceptions might have been different had they switched to careers to the arts, design, entertainment, sports, and media occupations. Determining the precise degree to which profitability is associated with quantities of digitized cultural capital or with authors' occupations is limited by the small sample used in this study.

PEDAGOGICAL IMPLICATIONS

In accounts of who publishes their résumés on the Web and finds such a Web venture symbolically profitable, a key factor seems to be the amount of digitizable cultural capital produced in one's occupation. Compared with generic one-page print résumés and their accompanying cover letters, the Web résumés and their accompanying sites studied here were large, evidence that their producers had substantial amounts of digitizable cultural capital to display.

Those in occupations that involve producing digitizable cultural capital were prominent within the sample not only in terms of sheer numbers but also in terms of the efficacy with which they rated their own Web résumés. Web résumés with more digitized cultural capital—including more graphics and being surrounded by more site files—tended also to be rated as more useful though this association might be a function of the occupations of those who post such digitized cultural capital and not simply of the sheer quantity of posted material.

By contrast, there was almost no representation in the sample from blue-collar occupations. Even white-collar occupations typically assumed to be well endowed with both cultural and economic capital were almost absent from the sample; their capital is presumably not the kind that would be readily digitizable for the Web market.

Those in the arts, design, entertainment, sports, and media occupational classification, which encompasses professional communication, were numerous and rated their Web résumés as relatively useful. The significant number of students in the sample may serve to remind us that, at least for select occupations, Web publishing is arguably perceived as one of the more accessible venues for showcasing a novice's digitizable cultural capital. Many students who have completed professional communication courses likely will have accumulated an impressive collection of digitizable cultural capital worthy of the Web résumé and portfolio project that has become a staple of the professional communication curriculum, as suggested by the attention the project has received in professional communication textbooks (e.g., Cunningham, Smith, & Pearsall, 2005; Markel, 2004).

For many professional communication students, however, a Web résumé and portfolio project presents considerable challenges. When I have taught such assignments, several students usually have struggled to find academic and professional documents worthy of publishing for the Web employment market and sometimes have resorted to publishing assignment drafts, test answers (without the corresponding questions!), obscure spreadsheets, and other documents that would no doubt puzzle anyone but the instructor who originally assigned them. Tellingly, in a recent undergraduate course in which my students could choose between creating a modest Web portfolio or a nondigital alternative, only a minority opted for the former. In a brief anonymous survey afterwards, about half the students indicated they felt they did not have material worthy of online display. Typical was the comment of a student who wrote, "[My] work experience wasn't sufficient enough to have a portfolio."

Such predicaments can also be common among professional communication students who enter the field from academic or professional backgrounds in which their work might not be readily digitizable or appropriate for the Web employment market. Even for students with considerable professional communication coursework, their academic programs conventionally help them develop skills in such areas as project management, collaborative work, and oral presentations; skills that, again, are not necessarily readily representable digitally. And the duration of a typical professional communication course does not always provide students with time to develop new, high-quality digitizable cultural capital for a Web portfolio.

These limitations in acquiring digitizable cultural capital—limitations arising from students' academic and professional backgrounds and compounded by the short duration of a one-semester course—can be as intractable as the limitations Bourdieu traced to family background and cultural capital's long gestation period (1986). Bourdieu recounts that his notion of cultural capital arose from observing that students' academic performances depended to an extent on social class, which endowed students with different types of cultural capital, in turn positioning them at different starting points in their academic development (p. 243). An academic Web résumé and portfolio project may risk reinscribing the class system when considerations such as digitizable cultural capital are not taken into account. Unlike other academic projects, which typically help students develop new cultural capital, an academic Web résumé and portfolio project often relies on students' already-accumulated endowments of that resource.

To counteract this heterophilic effect, a Web résumé and portfolio project could aim instead to develop new cultural capital in the form of two types of knowledge and skills:

- students' critical perspective on cultural capital itself, its development, and its role in their professional development and in employment markets; and
- students' abilities to communicate their cultural capital through the Web and to plan for further development of their cultural capital over the long term.

First, a Web résumé and portfolio project could aim to enhance students' critical understanding of how cultural capital develops in individuals and is converted, in employment markets, into economic capital. Such a perspective would enable students to step back from the résumé's traditional emphasis on individual enterprise and the employment markets' extension of this emphasis—competition among individuals. Instead, an understanding of Bourdieu's vision of the social cultivation of cultural capital might help students recognize that their own and their peers' embodied, objectified, and institutionalized cultural capital has developed not just through personal efforts but through support networks such as the family and larger society—in short, that the social endowment of capital, and not simply individual prowess, helps determine which positions workers occupy. This study has suggested that the position of Web résumé author seems open primarily to those whose occupations enable them to develop digitizable cultural capital. Students therefore could be encouraged to explore the social resources that have enabled them to develop their digitizable cultural capital, recognizing their academic professional communication projects in particular as opportunities for such development.

A critical perspective also would enable students to interrogate the basic formula that more is better: More Web pages, more Web graphics, and so forth. Whereas this study did find an association between more such digital content and greater perceived profitability, it is possible that other factors, such as an occupation that enables workers to readily develop digitizable cultural capital and a Web market enabling them to profit symbolically and financially from it, are responsible for some of that perceived profitability—not the sheer proliferation of digital content itself. Hence, the Web résumés and portfolios of students pursuing certain careers might not necessarily be made more profitable simply by adding more digital content because, for them, the Web market might not highly value such additional content.

As for students specifically pursuing professional communication careers, a critical perspective can create critical consumers, but a professional communication program must obviously also aim to develop students as effective producers. Despite the challenges of marshaling sufficient digitizable cultural capital to make a Web résumé and portfolio a symbolically profitable endeavor in the short term, such a well-supported project could nevertheless also foster planning for development of cultural capital over the long term. Students could begin their Web résumé and portfolio project, like most professional communication projects, with discovery activities, in this case taking inventory of their embodied, objectified, and institutionalized cultural capital. Following this process, students might sort out those kinds of cultural capital more readily digitizable and thereby representable on the Web and take inventory of what corresponding digitizable genres they either already have available to post or might develop throughout the semester. Readily representable on the Web are research, writing and design, artistic, and computer skills; and the corresponding digitizable genres include proposals, reports, manuals, brochures, and most other kinds of professional communication documents, including, of course, Web sites themselves. For cultural capital that is not readily digitizable, students could identify alternative communication channels as well as strategies for representing such capital, at least in part, on the Web. Managerial, collaborative, and presentational skills might be communicated more effectively through face-to-face interviews but could nevertheless be represented indirectly through the type of reflective statement that Miles Kimball recommends accompany a Web portfolio (2003, p. 22). If one of the portfolio's artifacts includes a collaboratively produced document, such a statement introducing the document could also characterize collaborative teamwork, explain the student's strategies for effective collaboration, and describe the skills that he or she developed as a result of the experience.

A Web résumé and portfolio project can cultivate long-term plans for developing cultural capital in additional ways. Given the obvious verbal, visual, and multimodal nature of work in professional communication, students could plot out those courses, degrees, or internships that would provide opportunities to develop digital resources. Capital-poor students could thereby supplement their modest accumulation of digitizable cultural capital by drawing on strategic thinking to create new work specifically to accompany their Web portfolio, such as a statement elaborating their career goals, their plan for attaining these goals, a personal statement of the ethics that guide their career development, and other capital-building planning documents.

This chapter has inquired into self-published Web résumés and has found a marked digital divergence favoring those whose occupations involve producing digitizable cultural capital. Drawing on Bourdieu's model of discourse production and forms of capital as well as on the survey responses of 100 Web résumé authors, I have argued that investing digitizable cultural capital in the Web market can be symbolically, and perhaps financially, profitable. Importantly, such capital is produced by professional communicators though students of the field may have only just begun accumulating their own digitizable cultural capital. Yet students can both develop a critical perspective on the role of cultural capital in their professional development and also plan to develop their digitizable cultural capital over the long term.

(*Appendixes follow*)

APPENDIX A
Number of Web Résumés by Standard Occupational Classification

Standard Occupational Classification[†]	*N*
Computer and Mathematical	39
Arts, Design, Entertainment, Sports, and Media	27
Management	9
Architecture and Engineering	9
Education, Training, and Library	5
Life, Physical, and Social Science	3
Business and Financial Operations	1
Healthcare Practitioners and Technical	1
Community and Social Services	1
Students	1
Legal	0
Healthcare Support	0
Protective Service	0
Food Preparation and Serving Related	0
Building and Grounds Cleaning and Maintenance	0
Personal Care and Service	0
Sales and Related	0
Office and Administrative Support	0
Farming, Fishing, and Forestry	0
Construction and Extraction	0
Installation, Maintenance, and Repair	0
Production	0
Transportation and Material Moving	0
Military Specific	0

[†]The Standard Occupational Classification was devised by the Bureau of Labor Statistics within the U.S. Department of Labor. Apart from the category of *students* added here, it consists of 23 major categories (see http://www.bls.gov/soc/soc_majo.htm).

APPENDIX B
Distribution of Participants by Age

Age	*N*
Under 20	1
20–29	37
30–39	37
40–49	18
50–59	3
60+	4

APPENDIX C
Median Numbers of Résumé Features and Differences between Rankings of Higher-Rated and Lower-Rated Résumés

Résumé Feature	Median of Higher-Rated Résumés	Median of Lower-Rated Résumés	$z^{\dagger}$
Site Files	178	75	2.16*
Internal Links	8	6	1.44
Graphics	5	2	3.10**

*$p < 0.05$. **$p < 0.01$.

$^{\dagger}z$ was tabulated for rankings using the Mann-Whitney *U* test, with adjustments for tied rankings using the formula in Hays (1981, p. 589).

APPENDIX D
Median Numbers of Résumé Features and Differences between Rankings of Arts, Design, Entertainment, Sports, and Media Résumés and of Computer and Mathematical Résumés

Résumé Feature	Median of Arts, Design, Entertainment, Sports, and Media Résumés	Median of Computer and Mathematical Résumés	$z^{\dagger}$
Site Files	211	87.5	1.39
Internal Links	11.5	12	0.46
Graphics	6	2	2.88**

**$p < 0.01$.

$^{\dagger}z$ was tabulated for rankings using the Mann-Whitney *U* test, with adjustments for tied rankings using the formula in Hays (1981, p. 589).

REFERENCES

Anderson, P. V. (2003). *Technical communication: A reader-centered approach* (5th ed.). Boston, MA: Thomson.

Bourdieu, P. (1977). The economics of linguistic exchanges. *Social Sciences Information, 16,* 645–668.

Bourdieu, P. (1986). The forms of capital ®. Nice, Trans.). In J. G. Richardson (Ed.), *Handbook of theory and research for the sociology of education* (pp. 241–258). New York: Greenwood Press.

Bourdieu, P. (1991). *Language and symbolic power*. (G. Raymond & M. Adamson, Trans.; J. B. Thompson, Ed.). Cambridge, MA: Harvard University Press.

Bureau of Labor Statistics. (2004, December). *Standard occupational classification (SOC) user guide*. Retrieved March 17, 2006, from http://www.bls.gov/soc/socguide.htm

Culwell-Block, B., & Sellers, J. A. (1994). Résumé content and format—Do the authorities agree? *The Bulletin of the Association for Business Communication, 57*(3), 27–30.

Cunningham, D. H., Smith, E. O., & Pearsall, T. E. (2005). *How to write for the world of work* (7th ed.). Boston, MA: Thomson.

Dominick, J. R. (1999). Who do you think you are? Personal home pages and self-presentation on the World Wide Web. *Journalism and Mass Communication Quarterly, 76,* 646–658.

Grabill, J. T. (2003). On divides and interfaces: Access, class, and computers. *Computers and Composition, 20,* 455–472.

Harcourt, J., & Krizan, A. C. (1989). A comparison of résumé content preferences of Fortune 500 personnel administrators and business instructors. *Journal of Business Communication, 26*(2), 177–190.

Hays, W. L. (1981). *Statistics* (3rd ed.). New York: Holt, Rinehart, and Winston.

Herring, S. C., Kouper, I., Scheidt, L. A., & Wright, E. L. (2004). Women and children last: The discursive construction of Weblogs. In L. J. Gurak, S. Antonijevic, L. Johnson, C. Ratliff, & J. Reyman (Eds.), *Into the blogosphere: Rhetoric, community, and culture of Weblogs*. Retrieved July 11, 2005, from http://blog.lib.umn.edu/blogosphere/women_and_children_pf.html

Hutchinson, K. L., & Brefka, D. S. (1997). Personnel administrators' preferences for résumé content: Ten years after. *Business Communication Quarterly, 60*(2), 67–75.

Killoran, J. (1999). *The virtual squatter: Homesteading in the electronic metropolis*. Doctoral dissertation, University of Waterloo, Ontario, Canada.

Kimball, M. A. (2003). *The Web portfolio guide: Creating electronic portfolios for the Web*. New York: Longman.

Lenhart, A., Horrigan, J., & Fallows, D. (2004). *Content creation online*. Retrieved March 2, 2004, from http://www.pewinternet.org/PPF/r/113/report_display.asp

Lenhart, A., & Madden, M. (2005). *Teen content creators and consumers*. Retrieved March 30, 2006, from http://www.pewinternet.org/PPF/r/166/report_display.asp

Madden, M. (2004). *Artists, musicians and the Internet*. Retrieved March 30, 2006, from http://www.pewinternet.org/pdfs/PIP_Artists.Musicians_Report.pdf

Markel, M. (2004). *Technical communication* (7th ed.). Boston, MA: Bedford/St. Martin's.

McDowell, E. E. (1987). Perceptions of the ideal cover letter and ideal résumé. *Journal of Technical Writing and Communication, 17*(2), 179–191.

Moody, J., Stewart, B., & Bolt-Lee, C. (2002). Showcasing the skilled business graduate: Expanding the tool kit. *Business Communication Quarterly, 65*(1), 21–36.

Oliu, W. E., Brusaw, C. T., & Alred, G. J. (2004). *Writing that works: Communicating effectively on the job* (8th ed.). Boston, MA: Bedford/St. Martin's.

Popken, R. (1999). The pedagogical dissemination of a genre: The résumé in American business discourse textbooks, 1914–1939. *JAC, 19*(1), 91–116.

Royal, C. (2005). A meta-analysis of journal articles intersecting issues of Internet and gender. *Journal of Technical Writing and Communication, 35*(4), 403–429.

Selfe, C. L. (1999). *Technology and literacy in the twenty-first century: The importance of paying attention*. Carbondale: Southern Illinois University Press.

U.S. National Telecommunications and Information Administration. (2004, September). *A nation online: Entering the broadband age*. Retrieved March 6, 2006, from http://www.ntia.doc.gov/reports/anol/NationOnlineBroadband04.pdf

Van Dijk, J. A. G. M. (2005). *The deepening divide: Inequality in the information society*. Thousand Oaks, CA: Sage.

http://dx.doi.org/10.2190/CWDC6

CHAPTER 6

A Case Study of the Impact of Digital Documentation on Professional Change: The WPA Electronic Mailing List, Knowledge Network, and Community Outreach

Huiling Ding

OVERVIEW

This chapter describes how the Writing Program Administrators electronic mailing list facilitates and documents disciplinary knowledge-making activities. Nine categories of such activities are identified and discussed, and the influence of the digital on professional outreach is examined. Of special interest is the move from print to digital documentation. Data from the study suggest that an electronic mailing list can create a supportive learning and participatory environment for professional researchers and practitioners and can enhance, through online community outreach, the influence of a profession.

INTRODUCTION: THE INTERNET, COMMUNICATION, AND WRITING

As individual and organizational communication has become increasingly mediated by technology, countless studies have examined the use of computer and Internet technologies in writing classrooms. But the impact of computer and Internet use on the construction of digital documentation within various professions or organizations has received little attention. That is, while teachers of computer-mediated communication stress the impact of the computer and the Internet on communicative practices, teachers of digital documentation focus more on the product of such communication, namely, on the writing produced and recorded. Exploration of the uses of technologies in the digital preservation of texts produced by disciplinary communities may shed light on the digital's influence on workplace practices, relationships among community members, and the evolution of the professions.

When it enables exchange of information, archival use of e-mail exchanges, and key-word searches of an archived database, an electronic mailing list can serve as a knowledge network promoting dynamic, participatory, and interactive communication among professional members. What is more, the archiving function enables the digital documentation of informal exchanges and discussions about issues highly relevant to a profession while the search function makes the content of the mailing list navigable and user friendly. Creating bridges among experts, semi-experts, and novices, such lists can offer professionals with various levels and types of expertise the opportunity to communicate with one another and to become involved in collaborative problem solving.

Since 1993, the Writing Program Administrators (WPA) has used an openly subscribed, fully archived, and powerfully searchable electronic mailing list for discussions about all aspects of WPA work; traffic volume has increased eightfold since the list began. Unlike the WPA's formal forums such as conferences and the *WPA Journal*, the mailing list is an informal online platform for discussion and documentation of a wide range of WPA issues that most likely would get lost in the traditional "format" of personal letters and corridor conversations because of the lack of official channels to preserve and to distribute such materials. In very rare instances, some subscribers choose to use pseudonyms and commercial e-mail addresses to make controversial claims without jeopardizing their professional reputations or relationships with peers, for subscribers' identities are not as fluid as those of anonymous users in online chat rooms. (The electronic mailing list operates with the assumption that its members will subscribe with their real names and the e-mail addresses of the institutions they work for.) Some subscribers may even come from fields outside of the WPA or composition studies.

This chapter analyzes the way in which the WPA electronic mailing list facilitates and documents disciplinary knowledge-making activities. I investigate and classify both the Council of WPA's use of its list to initiate and sustain conversations among members as well as the impact such use has had on the professional evolution of WPA as a dynamic, interactive, collaborative, and knowledge-intensive profession. I use this catalogued data to argue that the WPA electronic mailing list complicates views of WPA work as either administrative or intellectual and highlights the constant knowledge-making character of its work for both the community of WPAs and the wider community of compositionists. The notable content generated via the list identifies this technology-mediated communication as a valid platform for intellectual contribution and innovation, in doing so expanding concepts of scholarly knowledge making as formal only. Finally, I examine how, despite its potential to serve as a powerful vehicle for community outreach and public relations, an electronic mailing list requires careful planning and implementation in light of the needs of primary and other audiences to achieve these goals.

PROFESSIONAL HISTORY OF THE WPA

For many years, WPA work was widely misunderstood and undervalued as managerial work "that [did] not produce new knowledge and that neither require[d] nor demonstrate[d] scholarly expertise and disciplinary knowledge" (WPA Executive Committee, 1996, p. 92). Writing administrators, many of them untenured assistant professors, stood little chance of succeeding professionally because of their time-consuming administrative duties. But the 1992 manifesto *Portland Resolution,* which argued for the improvement of working conditions for WPAs, marks the beginning of the formulation of professional standards undertaken by the Council of WPA and the campaign for equitable treatment of WPAs in the academic reward system (Hult, Joliffe, Kelly, Mead, & Schuster, 1992). In 1997, the Council of WPA issued the publication "Evaluating the Intellectual Work of Writing Administration," a watershed document acknowledging and evaluating WPA work as intellectual rather than managerial. "Evaluating" also calls for the assessment of WPA work in terms of five distinct types of intellectual activity—program creation, curricular design, faculty development, program assessment, and program-related textual production—and provides assessment guidelines. Since the document's publication, the profession has been committed to the task of educating students, faculty, and administrators about the significance of WPA work and to developing models of assessment reflecting the complexity of WPA scholarship (Hult et al., 1992). As Shirley Rose and Irwin Weiser (2002) explain, "In the past several years, composition studies has begun to give increasing attention to acknowledging, describing, and valuing/evaluating the intellectual work of writing program administration, recognizing that WPAs play a critical role in the development as well as application of knowledge in the field" (p. 277).

Because the WPA electronic mailing list was initiated in 1993, one year after publication of the *Portland Resolution*, examination of its function as the informal forum and digital archive for WPA can illuminate the influence that digital documentation has had on professional change within WPA. Such documentation, when incorporating multiple perspectives and open access, can serve as a platform for professional change by enhancing virtual communication and collaboration and by serving as a textual repertoire for the arguments a profession makes. Allowing optimal participation from its subscribers, the WPA electronic mailing list documents the significance of work in the field of composition at a time when the council, concerned with the place of WPA in tenure decisions, has been calling for re-evaluation of WPA work as *intellectual* rather than solely *managerial*. This chapter demonstrates how WPA's use of an openly subscribed and fully archived electronic mailing list functioning as the knowledge network for the profession has facilitated transformation of that profession into a knowledge-intensive, collaborative community and, through community outreach, has contributed to the ongoing strategic argument about the value of the profession's work.

INTERACTIVE KNOWLEDGE NETWORK: FORMAL KNOWLEDGE VERSUS INFORMAL KNOWLEDGE

The misunderstanding that WPA members are administrators and not intellectuals creates a spurious binary between WPA and knowledge making, and rationalizes and perpetuates the lack of disciplinary acknowledgement of WPA work and the consequent marginalization of WPAs. This section will examine and deconstruct this binary, under the lens of Louise Phelps' (2002) continuum of knowledge making. The discussion of knowledge making will evoke a scholar going through the research process, which might entail formulating questions, completing a literature review, devising sound arguments, and publishing one polished, original piece of research in a national or regional academic journal, or presenting research findings in academic conferences. Such a definition of knowledge making, however, is narrow. It falls to one end of the continuum of knowledge making: Namely, to formal or explicit knowledge (Phelps, 2002, p. 9). But tacit/practical or informal knowledge, which "consists partly of technical skills, mental models, beliefs and perspectives so ingrained that we take them for granted and cannot easily articulate them," is also valuable (Nonaka, 1991, p. 98).

Chun Wei Choo's (1998) discussion of the three categories of knowledge in organizations further expands Phelps' continuum of informal knowledge. As Choo points out, an organization works with three categories of knowledge: Tacit knowledge, rule-based knowledge, and cultural knowledge (p. 11). *Tacit knowledge* (know-how, heuristics, or intuition) is implicit personal knowledge embedded in action and defies formulation or articulation. This type of knowledge is similar to the accumulated know-how possessed by expert practitioners and, insofar as it is difficult to externalize or to store systematically, takes years of observation, peripheral participation, and practice to acquire (Lave & Wenger, 1990). *Rule-based knowledge* refers to "explicit knowledge that is used to match actions to situations by invoking appropriate rules" (Choo, 1998, p. 12). *Cultural knowledge* refers to the organization's culture as communicated through oral and verbal narratives such as stories, metaphors, analogies, and mission statements (p. 12). In the case of the WPA electronic mailing list, tacit knowledge and cultural knowledge can be considered two parts of the informal knowledge produced and circulated among members.

Engaged in the practical and intellectual work of program administration, WPAs must possess and apply formal and informal knowledge to accomplish their daily work. Instead of venerating research as the only legitimate type of knowledge, WPAs use this complex knowledge-making process in all aspects of their work, such as research, pedagogy, teacher training, and assessment. Creation and documentation of both formal and informal knowledge takes place on the electronic mailing list constantly, which both provides evidence of WPA work as intellectual work and facilitates the expansion, development, and growth of WPA as a discipline.

THE FUNCTION OF THE ELECTRONIC MAILING LIST IN THE KNOWLEDGE-MAKING PROCESS

Because people with different professional experiences and levels of expertise subscribe to the list, it acts as an online discussion board for the entire community of practice of rhetoric and composition, and particularly for those experienced enough to become WPAs of composition programs. To some extent, the list serves as the unofficial space for distribution, circulation, and consumption of disciplinary knowledge and sometimes as the starting point for production of new scholarship.

In addition, as a forum for open information access and communication, the list provides the WPA community with opportunities for professional development through regular communication, reinforcement of strategic goals, and informal online training for less-experienced WPAs.

To offer insight into the ways in which the WPA electronic mailing list facilitates professional change, I closely followed the posts to the WPA electronic mailing list in the month of October 2005, compiled all posts with titles and actual messages into a corpus, and analyzed and categorized different types of knowledge-making processes taking place on the list and their influence on professional change. Using the constant comparative method described in Barney G. Glaser and Anselm Strauss' (1967) *The Discovery of Grounded Theory: Strategies for Qualitative Research*, I coded individual posts in the corpus into various categories as they emerged from the data. I then compared posts falling within the same category, noting both the properties of each category and their relations to other categories; subsequently, I decreased the original list of categories to those directly relevant to knowledge-making activities. Then I read the posts' titles for possible additional category suggestions and for properties of existing categories. I observed, altogether, nine categories of informal and formal knowledge-making activities in the list discussion. In the sections that follow, each category is defined and illustrated.

1. Circulation of Existing Formal Knowledge

At times, the WPA electronic mailing list has been used to circulate requests for recommendations of research resources such as books, articles, and software and to share important or interesting articles or databases. Use of the list as a reference generator or bibliographical machine is one of its most frequent functions. Given that a large proportion of WPA scholarship is published in books rather than in academic journals, that standard academic databases do not include book chapters in their search engines, and that disciplinary databases such as CompPile are updated rather infrequently, skillful users of research databases may miss significant newly published research. E-mailing the list thus becomes a reasonable strategy because the articles one fails to locate in research databases may be read or written by subscribers who can respond in a timely manner. Inquiries about research sources often serve as the starting point of collaborative brainstorming about specific research issues, which, after being archived digitally, can be researched and referred to by members interested in relevant issues. One instance was a post asking for references to publications dealing with "the role of emotion in the teaching and learning of composition," a query that attracted more than 20 responses pointing to sources from various disciplines.

Use of the electronic mailing list for such inquiries promotes an atmosphere of disciplinary collaboration in research and a sharing of sources, which in turn help members of the profession to conduct more scholarly research. The list thus may be especially helpful to subscribers working by themselves in small writing programs with limited support.

2. Solicitation and Documentation of Informal Knowledge such as Know-How or Personal Experience

One of the key arguments the WPA has been making about the profession is, as mentioned, the intellectual nature of WPA work. In "Evaluating the Intellectual Work of Writing Administrators" (1996), it is argued that writing-program administration should be refigured as "scholarly and intellectual work . . . worthy of tenure and promotion" because it "advances and enacts disciplinary knowledge" (p. 97). The most frequent use of the electronic mailing list is to solicit discussion about issues related to various aspects of WPA work such as pedagogy, curriculum development, teacher training, program assessment, funding, technology, placement, and collaboration with other programs. I will cite examples from existing posts to demonstrate different types of knowledge-making activities and the constant application of disciplinary knowledge by WPAs on a daily basis.

Responses from various institutions demonstrate the highly contextual nature of WPA work and the need for disciplinary knowledge by administrators attempting to fulfill their responsibilities. One thread in online freshman composition classes attracted a great deal of attention and was thoroughly discussed from the perspectives of pedagogy, the use of technology, instructor pay, teacher training, and assessment. For instance, one member shared his experiences using weekly synchronous class discussion and asynchronous posts and individual conferences with students on their final papers, in addition to his efforts to request research and development money. Because the teaching of online writing classes was still quite new within the field of first-year composition in 2005, teachers unfamiliar with online pedagogy learned from the practices of more knowledgeable teachers. In the instance of this post, as in that of many others, the list provided the WPA community with an asynchronous platform for discussion and exchange of program-related experiences and for collaborative problem solving, which could not have taken place on such a broad scale had they occurred in print or oral media. The electronic mailing list enabled discussion and informal research about emerging pedagogical issues that may yet become subjects for formal theoretical and pedagogical research in WPA work.

3. Discussion About New or Neglected Research/ Pedagogical Issues to Produce New Knowledge

When members of the field encounter new problems or come across neglected issues in research or pedagogy, they can use the electronic mailing list as a platform for collaborative brainstorming. Because it is difficult to find relevant publications when one conducts research about novel or little-studied issues, sending queries to the list and asking for suggestions and help may initiate a process of collective/ collaborative brainstorming and invention, which in turn may facilitate the research or problem-solving process. Posts falling into this category during the period studied are those about first-year online composition classes, teacher preparation for online writing courses, designing writing assignments, and pedagogical use of blogs. Such use of the list calls attention to new and emerging research or pedagogical issues, informs subscribers of relevant publications, and advances the understanding of practitioners in the field.

4. Solicitation of Input and Collaboration in Existing Research and Pedagogical Projects to Help Produce Knowledge

Some posts ask for assistance and participation in research projects, especially via surveys and questionnaires conducted by members of the disciplinary community. In most instances, those who have requested responses also promise to share findings with the community by means of different venues, including the list and forthcoming professional conferences. Because the list functions as a virtual space for communication, the distribution of surveys or calls can reach a large number of members of the discourse community, give rise to rapid and appropriate responses, and result in the timely sharing of research findings.

Moreover, with composition programs constantly influenced by local contexts and technological development, research can be undertaken on means of improving pedagogical practice as it pertains to these topics. As a followup to such research projects, posts to the list discuss potential ways of integrating findings into local contexts and of sharing alternative ways of improving teaching in writing classrooms, practices that may never appear in formal publications and may remain folklore for the community. Such posts produce, articulate, and circulate tacit/informal knowledge and inform subscribers about pedagogical practices in their colleagues' classrooms.

5. Solicitation of Input Solutions for Local Rhetorical Arguments: Articulation and Circulation of Tacit Knowledge

A regular part of WPA work is to make arguments about various aspects of local writing programs to secure adequate funding and support for these programs. Some posts ask for input to make effective arguments about the importance of composition courses, to defend reasonable teaching loads and reasonable pay for untenured instructors, and to differentiate writing courses from literature courses. One thread of discussion on possible ways to argue against "4-4 loads" pointed to existing publications on class size and composition workload and suggested possible ways of adapting such guidelines to local arguments. Because many subscribers may have gone through similar experiences and may have successfully made such arguments, such posts facilitate the articulation and analysis of tacit knowledge about similar rhetorical arguments, collaboratively create community knowledge, and serve as guidelines for future WPAs.

6. Platform for Participatory Discourses and for Different Types of Knowledge-Making Processes

One aspect that distinguishes the WPA electronic mailing list from other, more formal spaces for disciplinary discussion, such as conferences or publications, is the safe environment that it represents for less refined or polished ideas, its responsiveness to inquiries about theory and practice, and its shared appreciation for the messiness of practical knowledge for composition practitioners and WPAs—unlike print culture, which reveres scholarly publication and restricts the circulation of local documents. Put another way, the list does not marginalize teaching/administrative lore, failure narratives, or narratives that never reject the null hypothesis. It acknowledges the value of alternative discourses, allows multivocal conversations, and encourages community exchange and collaboration on all aspects of WPA theory and practice, thus supporting the ongoing argument about WPA work as intellectual by documenting its complicated nature. One example is a thread that discussed the Web 2.0 protocol Facebook and its possible pedagogical implications and applications. At a time when little research seemed available about the impact of social networking tools on writing classrooms, the posts helped fill in the gap by introducing and elaborating on issues such as power and authority, the influence of technology on students' lives, and the potential for advertising courses and reaching special populations for study. The thread also called attention to the need to explore the influence of new technologies and to incorporate such technologies in writing classrooms.

By eliciting and discussing various forms of knowledge and helping to develop them collaboratively, the list allows the discourse community to discuss issues without judging their immediate scholarly value. Gradually growing into a repertoire of practical and theoretical knowledge, the list functions as WPA's virtual knowledge network by allowing open information and communication access, enabling ongoing exchange and dialogue, empowering members with different types of knowledge, and transforming WPA into a more open profession.

7. Training Ground for Newcomers to the Field: Multilevel Mentoring in the Knowledge Network

Besides providing practitioners a forum for discussion and collaboration, the WPA electronic mailing list, by way of its powerful archiving functions, has become a valuable resource of professional information and knowledge. For newcomers to the field, such as graduate students or new WPAs, the electronic mailing list provides a comprehensive yet accessible library and a safe space for peripheral participation. Newcomers who feel intimidated by the possibility of raising previously discussed or naïve questions on the list can use the list's digital archive as a repertoire of practical suggestions and informal and formal knowledge about WPA. The list also facilitates the professional enculturation of apprentice WPAs by both exposing them to disciplinary discourses/practices and providing them with a low-risk platform by means of which they can choose to observe, selectively participate in, or immerse themselves in conversations. One thread, entitled "Online writing courses, and what kinds of students do best in them" asked for suggestions to improve students' success in online first-year composition classes. Responses offered not only theoretical and research references, but also practical pedagogical and curricular suggestions learned from previous online teaching experiences.

Experts in WPA in turn become mentors who use coaching and modeling to help newcomers acquire WPA theories and skills. As a result, the electronic mailing list, as WPA's knowledge center, helps both to transform WPA into and to present it as a scholarly collaborative profession. Although this topic was not addressed during the month I studied, on February 28, 2006, a thread titled "First time to CCCC? Yeah, me too" was initiated regarding the resources available for first-time attendees of the Conference on College Composition and Communication (CCCC), in which professors with multiple CCCC attendances offer suggestions about all aspects of the conference experience. Such suggestions included, but were not limited to, possible conference resources and activities for new attendees, strategies to choose the sessions to attend, and ways to decrease the costs of food, accommodation, and transportation. In this way, list members mentor newcomers regarding ways to learn in the conference and teach them professional survival and socializing skills, which together go far beyond the advice most students receive from advisors. The list also serves as a resource for future newcomers to the field, who may feel intimidated by the prospect of going to major conferences.

8. Forum for Social Apprenticeship for Practitioners: Collaborative Creation of Practical Know-How

Even though they are not current WPAs and may not work as WPAs in the future, many compositionists subscribe to the WPA mailing list simply to benefit from the tremendous information and constant exchange of knowledge taking place on the list. Many audience members are practitioners in writing classrooms who no longer have the luxury of formal or informal mentoring on pedagogy and teaching strategies. In fact, many professors protect their own teaching from the inquisitive eyes of colleagues because such privacy has long been considered an "academic freedom" highly valued by university professors. Because of its highly participatory and supportive environment, the WPA list provides a low-risk, low-pressure forum for the discussion of classroom issues and a means of coping with difficult situations.

For instance, one thread entitled "walking out of a class: what to do the next class" described a frustrating encounter that the author, a highly experienced writing professor, had had with a class that failed to do the required homework. Having left the room midclass because of students' lack of input, he asked the list for suggestions. The post prompted eight thoughtful, nonjudgmental responses offering a variety of strategies for dealing with similar situations, which, when compiled, could serve as a teaching scenario for a new-teacher practicum. For writing instructors working in small programs or in programs where mentoring is not sufficiently implemented, the list offers access to experienced and supportive peers who can respond quickly to requests for help by drawing from their own repertoire of know-how.

9. Creation and Storage of Knowledge About the Organization

One final function of the WPA electronic mailing list as a knowledge-making tool is that of creating and storing cultural knowledge about WPA as a profession, for both the organization and the wider rhetoric and composition community. Verbal narratives about the profession have been created in the forms of stories, metaphors, and mission statements that are circulated widely through the list. For instance, the argument about WPA as intellectual work has been reinforced through regular discussions about promotion and tenure issues as well as arguments about working conditions, including class size and instructor pay. Recently, another important discussion has occurred opposing the use of the SAT/ACT Writing Exam's one-draft timed writing sample for placement in college writing courses, because of its inability to reflect the test taker's ability to produce writing under real-world circumstances (NPR, 2005). This protest reflects and reinforces WPA's stress on the writing process, multidrafting, and revision, and conveys the group's message to a wider public.

As suggested by the nine categories described here, the WPA electronic mailing list offers a platform for sharing formal and informal knowledge among its communities and for creating and preserving knowledge-making activities that reflect the intellectual aspect of WPA work. The list therefore offers solid evidence in support of the ongoing argument that WPA work is not solely administrative. In addition to knowledge creating and sharing, the list functions as an outreach tool with which WPAs can influence not only compositionists but instructors in technical writing and other fields who are interested in incorporating writing into their curricula.

COMMUNITY OUTREACH

Although the community of writing programs (including first-year composition and professional/technical writing) has several electronic mailing lists, the WPA, compared with other lists such as ATTW, TechRhet, or H-Rhetor, enjoys the most active participation, the largest number of subscribers, and the most extensive digital archival and documentation practices. To some extent, the WPA list serves not only WPAs and graduate WPAs, but reaches out and functions as the forum for the entire community of composition studies, since many compositionists subscribe to the list even though they are not WPAs.

Serving the Larger Community of Compositionists

By involving compositionists in the conversation, the WPA list bridges the gap between administrators and practitioners, creates and maintains connections among administrators and practitioners in composition programs, and helps distribute and reinforce the rhetorical argument about WPA as knowledge-making work for practitioners in composition studies. Such community outreach helps educate less experienced members of the field of composition studies about the work done by WPAs, the intellectual character of its work, and tested strategies for enhancing research and teaching. For instance, one thread discussed difficulties arising when a parent wrote an assignment for a student. Subject headers for the responses to the issue used different titles, such as "parents writing papers and other naive adults" and "my students don't plagiarize." The issue was discussed from the perspective of college and high-school writing instructors, classroom regulations, parent involvement in student writing education, and the use of Web sites such as Turnitin.com. The discussion turned the topic of plagiarism into an issue about classroom practices and strategies to deal with parents and students, a serious concern for composition instructors, and helped enhance understanding of both the scale of these behaviors and the motivations behind them. Such discussions strengthen the argument that WPA work is intellectual in that it involves the application of disciplinary knowledge to the solution of specific problems.

Involving Teachers in 2-Year Colleges

Another strength of the WPA electronic mailing list in terms of professional outreach lies in its open subscription, which invites and enables either passive observation or active participation from compositionists or WPAs in 2-year colleges. Because most 2-year colleges stress teaching exclusively, their composition instructors may do very little research and may participate only rarely in academic conferences such as WPA or CCCC due to their heavy teaching load and lack of research funding (Hutton & Jobe, 1985). As a result, it can be very difficult to get teachers at 2-year institutions to become involved in disciplinary research, journals, or conferences although the wisdom and expertise they can bring to the table is often substantial. Many compositionists in 2-year colleges do subscribe to the WPA electronic mailing list and participate in its discussions. Given the fact that increasing numbers of students take composition courses from community colleges to save educational expenses, subscription to the WPA list can help community college WPAs and compositionists to understand the WPA's professional mission and to utilize scholarship and research findings to improve pedagogical practices. The WPA list can also contribute to the professional development of compositionists and WPAs in 2-year colleges by offering them an informal forum in which to discuss all aspects of their work.

Possible Strategies to Further Expand Community Outreach

From time to time, members will post issues about professional and technical writing and writing across the curriculum (WAC) to the WPA list and will ask for help in making pedagogical improvement or local arguments about class sizes and teaching loads. The WPA list ultimately fails, however, to reach out to and to engage writing instructors from related disciplines in the ongoing conversation about writing pedagogy, research, and administration. Despite its advantages, the professional segmentation of the field of writing studies into first-year composition, professional/technical writing, WAC, and writing in the disciplines (WID) creates barriers against conversation among members of these subareas of writing studies. The WPA can do more to expose colleagues across English studies to the successful rhetorical arguments it already has made about teaching loads, class sizes, and teacher pay and promotion for composition classes, thus initiating cross-fertilization in both pedagogy and scholarship. Possible approaches to more actively involving instructors of professional writing, WAC, and WID are to solicit greater participation from

WPAs who also supervise professional writing, WAC, or WID programs and cross-posting related issues to electronic mailing lists such as ATTW or TechRhet for more involvement from those programs. Such outreach to subareas of writing studies can both expand the influence of the WPA as a professional organization and exert intellectual influence on similar arguments made in those subareas.

Such cross-disciplinary conversation, however, should be initiated without any "colonizing desire" (Bergmann, 1998); that is, WPAs should attempt to participate in mutual learning and to avoid unidirectional pronouncements about the right way to teach writing. Effective exchange and collaboration between the WPA and other writing-related disciplines can rely on the respect of "academic pluralism" and "disciplinary values or epistemologies," as well as on the preservation of consensus and maintenance of compromise, to the end of serving students (p. 154).

In addition, it must be acknowledged that the presence of WPA in a public space, especially that outside the university, is still invisible. The list, like other channels of communication for WPAs, is limited within the community of WPAs and compositionists and does not involve students' parents, university administrators, or public decision makers. To some extent, the communication channel between the public, the university, and WPAs is a closed circle: WPAs report to deans of liberal arts colleges, and deans offer them administrative directives. Meanwhile, the electronic mailing list's aggressive Internet presence, flexibility, and responsiveness presents the Council of WPA professionals as supportive, open minded, dynamic, and technologically savvy and greatly enhances their visibility within the wider community of writing instructors. Future studies should examine whether the WPA electronic mailing list can function as a bridge between the profession of WPA and the wider public and how the list might be used to achieve that purpose.

IMPLICATIONS

Electronic mailing lists can have a significant impact on professional productivity and innovation when implemented effectively, as demonstrated in the case of the WPA, whose list both strengthens the council's argument that WPA work is intellectual and breaks down the artificial divide between administrative and intellectual work. Through discussion threads, the list explicitly demonstrates the knowledge-making nature of WPA work, allows virtual collaboration and multivocal conversations among WPAs with different levels of expertise, makes connections between existing WPA theory and local practices, articulates tacit or informal knowledge in problem-solving processes, and provides an informal knowledge reservoir and a set of training materials to existing and new WPAs. Moreover, community outreach enabled by the WPA electronic mailing list demonstrates its potential to bridge the gap between administrators and practitioners through collaborative creation and dissemination of knowledge shared on a daily basis.

My analysis of the function of the WPA electronic mailing list as a platform for informal knowledge making and community outreach is far from comprehensive and no doubt omits important connections between such lists and the ways in which various professions may have put them to use. Clearly, however, when used as a virtual knowledge network for the community, an electronic mailing list can create a supportive learning and participation environment for professional researchers and practitioners, facilitate key knowledge-making endeavors of that profession, and enhance the influence of the profession through community outreach.

REFERENCES

Bergmann, L. (1998). Missionary projects and anthropological accounts: Ethics and conflict in writing across the curriculum. In S. L. Fontaine & S. Hunter (Eds.), *Foregrounding ethical awareness in composition and English studies* (pp. 144–159). Portsmouth, NH: Heinemann.

Choo, C. W. (1998). *Information management for the intelligent organization: The art of scanning the environment.* Medford, NJ: Information Today, Inc.

Glaser, B., & Strauss, A. L. (1967). *The discovery of grounded theory: Strategies for qualitative research.* Chicago, IL: Aldine De Gruyter.

Hult, C., Joliffe, D., Kelly, K., Mead, D., & Schuster, C. (1992). The Portland resolution: Guidelines for writing program administrator positions. *WPA: Writing Program Administration, 16*(1–2), 88–94.

Hutton, J., & Jobe, M. (1985). Job satisfaction of community college faculty. *Community/Junior College Quarterly of Research and Practice, 9*(4), 317–324.

Lave, J., & Wenger, E. (1990). *Situated learning: Legitimate peripheral participation.* Cambridge, UK: Cambridge University Press.

Nonaka, I. (1991). Knowledge-creating company. *Harvard Business Review, 69*, 6–104.

NPR.org. (2005). *The SAT essay.* Retrieved on April 18, 2008, from http://www.npr.org/takingissue/20050309_takingissue_sat.html

Phelps, L. W. (2002). Turtles all the way down: Educating academic leaders. In S. Brown & T. Enos (Eds.), *The writing program administrator's resource: A guide to reflective institutional practice* (pp. 3–40). Mahwah, NJ: Lawrence Erlbaum.

Rose, S., & Weiser, I. (2002). The WPA as researcher and archivist. In S. Brown & T. Enos (Eds.), *The writing program administrator's resource: A guide to reflective institutional practice* (pp. 275–290). Mahwah, NJ: Lawrence Erlbaum.

WPA Executive Committee. (1996). Evaluating the intellectual work of writing program administrators: A draft. *WPA: Writing Program Administration, 20*(1–2), 92–103.

PART III

Building Communities

http://dx.doi.org/10.2190/CWDC7

CHAPTER 7

A South-North Online Collaboration between Professional Writing Students in Tunisia and the United States

Faiza Derbel and Anne R. Richards

OVERVIEW

Following is an in-depth look at the potential of online collaborations connecting professional communication classrooms from the technologically enthusiastic South with those of the technologically accomplished North. By dealing more or less exclusively with others who are more or less like oneself, it is impossible to grasp the core realities of globalization, under which individuals of vastly different means and mentalities work and struggle. But in South-North online collaborations, the realities of globalization have a greater potential to surface. This chapter analyzes experiences with differences in computer access and literacy, and accounts of divergences and improvised solutions, of U.S. students of technical communication and Tunisian students of business English. The procedural, introspective, and retrospective data making such thick description possible serve as grounds for recommendations concerning cross-cultural online collaborations in highly heterogeneous educational settings in a postcolonial and globalizing era.

Advances in computer technology and increasing global access to the Internet are suggesting novel classroom approaches for professional communication teachers around the world. Both a society's need to educate informed and thoughtful members and a student's need to find gainful employment in a globalizing workforce characterized by the growing presence of transnational distributed work can be addressed through collaborative online projects with a focus on comparative digital cultures.

Online cross-cultural collaborations tend to be undertaken, however, between classrooms in the comparatively affluent North, where computer-mediated pedagogies are more familiar to students and more thoroughly integrated across curricula.

Because these collaborations always require a considerable commitment beyond a teacher's regular responsibilities, the added complication of significant digital divergences in computer access and literacy that exist between classrooms of very disparate economic means may threaten to undermine the teacher's ability to cope. All too aware of such difficulties, we nonetheless argue in this chapter that online collaborations between the technologically accomplished North and the technologically enthusiastic South are not only feasible but uniquely valuable because they have the potential to illuminate a vast range of digital landscapes and to help participants identify and shake off tenacious cultural stereotypes. Critiquing professional communication's "sociopolitical functions and effects, including how they shape people's subjectivities and material experiences" and "strategiz[ing] ways to intervene and transform hegemonic forms of power," this chapter engages in the cultural turn advocated by J. Blake Scott and Bernadette Longo in the introduction to their 2006 special issue of *Technical Communication Quarterly* (pp. 3, 4).

The elephant in our room is globalization, which we have conceptualized linguistically and discursively as a "transcultural postmodern condition whereby ideologies continuously flow between the culturally dominant and less dominant, though more insistently from former to latter, and are conveyed primarily by means of English in any of its countless world varieties" (Derbel & Richards, 2007, p. 4). The language of instruction and of cross-cultural communication in the collaboration we describe in this chapter was, as typically happens in the "real" world of global trade, the native language of the technologically more accomplished partners—in this case, English. Thus, the characteristic monolingualism of the U.S. students, which demanded that all cross-cultural clarification take place in their native language, contributed to the unequal flow of communication between classrooms, ironically to the advantage of the linguistically less accomplished group.

In the cross-cultural online learning space that we constructed with our students, issues of divergence in computer access and literacy were also foregrounded. One of the key goals of this chapter is to sketch digital technology's role as an artery of globalization. In conjunction with world Englishes and global capitalism (Giddens, 2000), digital technologies simultaneously promote cultural standardization and offer to many a means of economic and civic advancement; a concurrence suggesting the frustratingly ambiguous timbre of globalization. As Timothy Allen Jackson (2004) notes regarding the impact of new media, for instance,

> Although new media offers new possibilities and sites for transforming the power relationships of information production and dissemination, it simultaneously provides new means of reproducing dominant subject positions and power relations. Within identity politics, new media can function to both enhance, restrict, or enhance and restrict, the growth of critical consciousness. (p. 289)

Bearing in mind the effacing effects of digital technology, global capitalism, and the ascendancy of English as a world language (Crystal, 1997), we nonetheless recognize a hint of a possibility of developing more accurate and respectful notions of other cultures through these same vehicles (Rickly, 2004).

We have found that online cross-cultural collaborations between English-language professional communication classrooms have the potential to address the forces we have just identified (see Table 1). But these collaborations must be more than mere exercises in distributing power rationally or of discovering how the materially and technologically privileged can most efficiently organize and put to use the labor of the other. Collaborations of the type we advocate must insist on exploring alterity and ethics in the context of globalization. Given the problematic character of globalization, collaborations espousing the goals of critical pedagogy might help students develop empathy, appreciation for alternative models of viewing world civilizations, and facility for detecting the presence of various models in the classroom and beyond. Successful collaborations should also help students from the technologically accomplished North develop respect for the achievements of their partners, including their linguistic achievements as speakers of English as a second or other language (see Derbel & Richards, 2004). In the broadest terms, successful South-North online collaborations will help students cultivate what Michael Byram (1997) refers to as intercultural communicative competence, or the ability to "trade places" with others and to interact with high levels of cultural awareness and increased knowledge of culturally specific phenomena. Because the material conditions of our own classrooms were so different, students' communicative competences were mediated by digital technology not only as a tool but as a topos; that is, their cultural awareness and knowledge of culturally specific phenomena, including their own practices, were transformed by what they learned about their partners' relations to the digital.

Following is a brief discussion of relevant literature, as well as overviews of the online collaboration between our classrooms and our respective students' expectations concerning their international peers. The chapter's primary focus is the influence of digital culture on students' attitudes toward and participation in the collaboration. We conclude with a reflection on rationales and best practices in cross-cultural online collaborations between technologically very heterogeneous classrooms.

Table 1. Pedagogical Engagement with the Engines of Globalization. South-North Online Collaborations in Professional Communication Classrooms have the Potential to Engage All Three Engines Simultaneously, i.e., to Model Globalization, for Good or Ill.

Engine of globalization	Relevant classroom strategies
Digitalization	Online collaborations
Global capitalism	Projects in business or technical communication
English as a world language	English-mediated interactions between native speakers of English and speakers of English as a second or other language—or between speakers of English as a second or other language

POTENTIALS OF CROSS-CULTURAL ONLINE COLLABORATIONS

According to the United Nations, the use of computer technology in educational settings not only is radically changing the learning process but also is bringing about the restructuring of school as a learning space (UNESCO, 2005). Definitions of *literacy, communicative competence,* (Byram, 1997; Cope & Kalantzis, 2000; Rassool, 1999), and *effective communication across cultures* (Cameron, 2002) are being reframed as teachers around the globe structure collaborations between their students and others in different classrooms, schools, and nations. Teachers have pursued a variety of goals in the online collaborations we are familiar with, including those of helping students

- to practice using computer technologies for e-mail and online research,
- to engage in collaborative work with peers at home and abroad,
- to write for a "real" international audience,
- to practice communicating with speakers of the target language, including native speakers,
- to engage in self-reflection and peer learning, and
- to learn to detect others' perspectives and to reconstruct their own through sharing experience and conversation across curricula and cultures.

Many researchers have described collaborations involving U.S. students and peers from countries across the Atlantic (e.g., Belz, 2003; Kern, Ware, & Warshauer, 2004; Kramsch & Thorne, 2002; Yowell, 1995), as well as English as a Second Language (ESL) and/or English as a Foreign Language learners, or learners of languages other than English (O'Dowd, 2003). Length and focus of collaborations have depended on teachers' personal interests and curricular goals and on technological capabilities (Fedderhöldt, 2001; Kaufman, 1998; McClanahan, 2001; Yowell, 1995). Clearly, the potential to engage international peers online is very great in the United States. Between 2002 and 2007, the number of postsecondary students taking at least one online course more than doubled, from just over 1.6 million to nearly 4 million. This shift reflects an increase in online enrollment, which as a percentage of total enrollment increased from nearly 10% to more than 20% over the same period (Allen & Seaman, 2008).

What is more, Internet access has been increasing rapidly around the globe (see this collection's introductory chapter for a discussion of this phenomenon), and although the literature on online collaborations involving the outer and expanding circles of English language use has been thin so far, academic collaborations involving the economic South are surely increasing. Research on cross-cultural online collaborations has tended to describe North-North encounters for a number of reasons, including the North's comparatively rapid embrace of digitally informed pedagogy, which likely reflects not only technological capacity but also classroom teaching style. The clear exception has been China (e.g., Dautermann, 2005), which figures in a number of studies of online collaborations between technologically very heterogeneous classrooms in English studies and other fields. The reasons for China's rapid inclusion on the list of desirable online partners is obvious, given that China's GDP in terms of purchasing power is the third largest in the world, following the GDPs of the European Union and the United States, in that order. In contrast, Tunisia's GDP is ranked 73rd globally out of a field of 266 nations (CIA World Factbook, 2009b).

AN E-COLLABORATION BETWEEN TUNISIAN AND U.S. CLASSROOMS

Background

Communication with individuals outside the national borders of our respective classrooms was a goal we both cherished as a result of our experiences as teachers of English as a second language and of multiculturalism. As graduate students within the same department on September 11, 2001, we found ourselves mutually drawn to the advantages of cross-cultural communication as a way of helping students to understand the situations and aspirations of their peers around the world. After graduating, Derbel returned to her home country of Tunisia and accepted a position as assistant professor with a university in that country's second business, industrial, and technology hub, while Richards continued to teach for the following year at their graduate institution.

Objectives

The mutual overarching goals of the collaboration put into practice in spring 2004 were to encourage students to communicate with others outside their national borders in order to develop as professional writers and to achieve a clearer understanding of their own and their partners' cultures. The primary goals for our respective students were as follows:

- The U.S. students were expected to write for a specific international audience and to revise their work according to suggestions provided by that audience, and
- The Tunisian students were expected to practice communicating with native speakers of English and to consult with them on matters of e-commerce.

Inspired by the writings of Kramsch and Thorne (2002), Fürstenberg, Level, English, and Maillet (2001), and Warshauer and Kern (2000), to name only a few, Derbel saw in online collaboration an opportunity for learners to improve their language skills through interaction with native speakers of English. She saw the collaboration as a context for her students to experiment through participation and to see for themselves how they were faring, by noticing the effects of their communicative choices on their interlocutors. Moreover, she held the view that collaborative work with peers living in another cultural setting has the potential to encourage reflection and to enhance one's ability to detect others' and to revise one's own perspectives (Byram & Fleming, 1998; Knowles & Hennequin, 2004; Kramsch, 1993; Rickly, 2004). Consequently, she hoped to create, through the collaboration, a learning environment favorable for the generation of her students' own ideas and for appreciation of unfamiliar ideas held by their U.S. partners. Specifically, she expected her students to articulate as part of the participatory process their own ideas and to develop voice in the English language. She believed that part of her students' preparation for future careers in business in Tunisia (or abroad) would be served by infusing technology deeply into their program of study. The e-commerce course she was teaching lent itself to the use of computers and to research on the World Wide Web. In short, Derbel's reason for involving her students in online collaboration was twofold: On the one hand, to develop their language, communication, and interpersonal skills and, on the other, to improve their technological literacy skills (Cope & Kalantzis, 2000).

Richards shared most of Derbel's key goals, including those of facilitating intercultural awareness and intercultural peer-to-peer feedback in a comparatively "authentic" setting. She also expected her students to demonstrate by the project's close a broad knowledge of Tunisia as a location for technical writing in English; to practice Internet research skills relevant to that international setting; and to demonstrate awareness, through self-reflective writing and informed discussion, of Tunisia's postcolonial and globalizing present.

Course Descriptions and Participants

The Tunisian group included 26 students enrolled in a junior-year undergraduate course in business English, which weekly entailed a lecture and a 2-hour practice session in a computer laboratory with Internet connection. The main challenge for the Tunisian students was the lack of examples of e-commerce practice, for this remains an emergent aspect of Tunisian economic life. During their sessions, students learned how to search the Web for business and e-commerce-related information and to explore, browse, and analyze the sites of Tunisian, U.S., and other international companies engaged in e-commerce, a comparatively novel phenomenon in Tunisia at the time, where, in spring 2004, credit card usage was being restricted due to fiscal policies aimed at controlling the value of the dinar. In contrast to those of their U.S. counterparts, the Tunisian students' final grades in the course did not depend on completion of tasks and assignments. Instead, students sat for a high-stakes comprehensive exam at the end of the semester, in this instance on topics related to business English in theory and practice.

The 22 U.S. students were enrolled in a 300-level undergraduate technical writing course in a Midwestern land-grant university of science and technology. The class met 3 hours a week, 1 hour of which was in a fully wired and otherwise state-of-the-art computer laboratory, usually attended by a monitor who, along with the teacher, helped students navigate the stages of computer-mediated learning. Many of the students were computer-science majors for whom technical writing was a requirement for graduation. *Technical Communication*, by Rebecca Burnett (2000), and *Global Contexts: Case Studies in International Technical Communication*, by Deborah S. Bosley (2000), were the primary reading materials for the course. Students were expected to develop fluency in WebCT, the Macromedia software Dreamweaver, and various e-mail programs, and to post to WebCT responses to teacher prompts that at times directly concerned the progress of the collaboration. Students also were assigned three group portfolios, one of which was to be revised in light of correspondence with Tunisian students.

Because e-mail was by far the dominant mode of asynchronous computer-mediated communication at the time, we created and matched the cross-cultural groups and guided our respective students in communicating with each other by that means. We hoped that our teaching roles would be limited mainly to setting up the conditions for learning and that students on both sides would be active, uninhibited communicators.

Methods

Most of the online collaboration research we are aware of that reflects a second language acquisition (SLA) perspective has studied learner output resulting from group interaction, through the use of posttests (Kitade, 2008) and quantitative (sentence length) and qualitative (sentence complexity) (Fedderhöldt, 2001; Kaufman, 1998; McClanahan, 2001) discourse analyses. Studies undertaken from an interactionist SLA perspective, for instance, have attempted to gauge the impact of technology use on the patterns of interaction occurring during the collaboration (Kern, 1995; Wang, 2006) and, more specifically, on learners' use of communication strategies such as self-repair (Smith, 2008) and peer feedback (Lee, 2006). In contrast, researchers involved in the implementation of high-profile collaborative projects (Belz, 2003; Fürstenberg et al., 2001; O'Dowd, 2003) have tended to function in a more holistic framework and to explore themes such as students' use of selected instructional input as an attempt to assess the contribution of input and the exchange of ideas during online collaboration, with the development of "intercultural competence" (Liaw, 2006) and the tracing of attitude change (Kramsch & Thorne, 2002) as a target outcomes. These reports have identified and discussed difficulties associated with the implementation of projects and pinned down difficulties arising out of cross-cultural communication during collaborative work involving students from different countries (Belz, 2003; Kramsch & Thorne, 2002; Lee, 2004; O'Dowd, 2003).

From a professional communication standpoint, R. Peter Hunsinger (2006) observes that much research on intercultural communication has tended to take a heuristic approach by positing and analyzing binary cultural traits such as individualism/collectivism, high context/low context, or masculinity/femininity. Hunsinger finds this framework constraining and offers what he describes as a "transtextual" framework in which the notion of cultural identity is located alongside themes key to understanding globalization, such as economy, politics, history (p. 36). He argues for the usefulness of Arjun Appadurai's five global flows of textuality (ethnoscapes, mediascapes, technoscapes, financscapes, and ideoscapes, p. 40), observing that these themes can help generate especially useful insights into intercultural communication because they insist that cultural identity is fluid and emerges in the context of globalization.

Specializations within English studies clearly diverge in terms of the values they assign specific research methodologies. As practitioners in very different subspecialties of English (Derbel in linguistics and Richards in rhetoric and professional communication), we nonetheless concur that there is a place for quantitative methods and for objectivist assumptions in the field of professional communication. For all we know, there may be aspects of culture that are both influential in the moment and relatively stable (aspects that will outlast the digital era as well as the era of globalization), and if so, recognizing these aspects can be of use to professional communicators. Yet we also appreciate Hunsinger's politically attuned qualitative approach, which encourages us to explore culture in light of some of the central tropes of postmodernity. Our own research is of this second type, focusing as it does on the technological relationships we forged with our students while fulfilling the collaboration's goals.

Our analysis of the collaboration centered on these two research questions:

1. What expectations did students have concerning intercultural online collaborations?
2. How did divergences in technological environments influence students' attitudes toward and participation in their collaboration?

Because the collaboration was assigned to students taking different courses and assigned different tasks, involved both native speakers of English and speakers of English as a second or other language, and took place in different educational systems and distant countries, and because we had no joint access to computer software that would have helped us answer our research questions, we elicited data by those methods we felt were most appropriate to our complex research situation. The U.S. technical communication students were writing extensively on WebCT about their collaborative experiences, and we studied this group by gathering and analyzing the content of their electronic texts as well as their relevant portfolios. Because ongoing assessment was not possible in the Tunisian system and we felt it would have been unfair to divert substantial class time from the topic of the exam, we had no data log to fall back on. As a result, we resorted to retrospective interviews with available participants at the end of the semester. During the interviews, we asked students to discuss their expectations of the project, their perceptions of its usefulness, and the experiences they had had with computer technology.

Because they were writing about their online experiences on WebCT and in the self-reflection component of their final portfolio, we had more and richer data to work with from the U.S. students. Again, unlike with the U.S. curriculum, the Tunisian curriculum could not be adjusted to give students credit for participating in the collaboration. As has been mentioned and as will become even clearer, the responsibility for participation in the research was just one example in which the burden of collaboration fell more heavily on the Tunisian collaborators.

STUDENT EXPECTATIONS

Synthesizing the ideas that students gave for their expectations of the project and of their peers, Tables 2 and 3 suggest, among other things, that although teachers create the context for online collaborations, students interpret the purposes of these projects according to their own frames of reference.

The interviews reveal no instances in which the Tunisian students expressed concern over whether U.S. citizens in general or more specifically their U.S. peers might stereotype or have negative expectations of Tunisians, nor did the Tunisian students anticipate any of the technological difficulties they were to encounter during the course of the project. The Tunisian students indicated, in short, having had mainly positive expectations of the collaboration. Not only had they been excited by the prospect of interacting with U.S. students, but some had constructed rather elaborate academic goals that extended beyond the semester. One of the advanced students, Darine (the names of all students have been changed), felt "it would be interesting to have this communication and this exchange of ideas, this collaboration, for two years, and why not longer? Why not? There would be many things to do." Clearly, she had very high aims for the project:

> I expected at the end that we could have an exchange of places. They could visit Tunisia, and we could visit America—we could have a project that we

Table 2. Students' Stated Positive Expectations of the Online Collaboration

Rationales	U.S. students	Tunisian students
Affective	Discovering new things	Developing personal ties and creating long-lasting friendships with Americans that would lead to future projects and perhaps other types of connections
	Doing a challenging project	Discovering what "real" Americans are like
Instrumental	Learning about education abroad	
	Learning about cross-cultural communication	
	Practicing cross-cultural communication	
	Learning from Tunisian students	
Ethical	Being looked to as experts	
	Being cultural ambassadors	

Table 3. Students' Stated Negative Expectations of the Online Collaboration

Rationales	U.S. students	Tunisian students
Affective	Giving offense	The Tunisian students did not voice concern that the U.S. students would/could hold negative views of them as Arabs or Muslims or as inhabitants of the economically developing world. They did not anticipate difficulties communicating through technological means, nor did they anticipate difficulties in communicating in writing or in English.
Instrumental	Experiencing slowed communication and completion of final project Experiencing the technological divide	
Ethical	Dealing with anti-American stereotypes Dealing with inaccurate perceptions of cultural norms Dealing with deficient language use on the part of the Tunisians Being ineffective writers	

> create together. . . . I expected that we could, at least, do two-week internships in America. It was really bad to find out that it was only e-mail exchanges—questions, answers, I don't know. At first I felt it was a big project. Maybe because it was my first time.

This same student also had in mind developing long-lasting personal ties, holding joint conferences, and hosting U.S. lecturers. Echoing Darine, Siwar commented, "I thought that the collaboration will finish with a conference, a project, something like that. But I was surprised when we finished without . . . like this . . . anything. Surprised [not disappointed]."

Although the Tunisian students were conscious of the fact that this was an exchange carried out in an academic context, some had tried, and others had wished that it were possible, to create lasting friendships with their U.S. peers. "Even when we finished our academic year, I wanted to continue the collaboration," Siwar told us. "But I felt very uh [awkward] . . . maybe it was meant only for studies . . . that's why I didn't continue." That certain students from both classrooms did indeed consider the possibility of remaining in contact after the semester strengthened our belief in the value of the collaboration though the quality and degree of interest demonstrated suggest that the U.S. students had a more instrumental attitude toward the project and toward their partners than the Tunisian students did. This may be so because of the nature of assignment expectations and assessment systems, as described earlier.

From the beginning, in fact, the Tunisians were looking forward to communicating with their U.S. peers in order to discover what "real Americans" were like and how they lived, studied, and worked. Darine felt that such knowledge would be important, for Tunisians, in her view, were often exposed to stereotypical images of the United States:

> I was interested to know what Americans—the young generation (the teenagers)—how they think. Sometimes the idea we get about Americans . . . the idea sometimes is not objective. . . . The media sometimes pass negative ideas about Americans. Sometimes it's good . . . so I am fascinated by their way of life.

Excited and strongly motivated by the novelty of the project, on commencing with it, the Tunisian students seemed to have been unconcerned with potential difficulties.

In contrast, although the U.S. students voiced a number of positive expectations regarding the collaboration, they voiced a number of negative expectations regarding their international peers. The negative expectations most commonly mentioned by the U.S. students were that stereotypes, inaccurate perceptions of cultural norms, and language and religious differences would interfere. The U.S. students were also preoccupied with the possibility that they would unwittingly offend their international collaborators or that technology would fail on the Tunisian side.

The U.S. students seemed to be more conscious than the Tunisian students about stereotypes generated out of the context of political conflicts and were not fully aware that Tunisians, although 98% are Muslims, are geographically more distant than, say, Syrian students from the zone of conflict in the Middle East. The U.S. students were very worried that the Tunisians had stereotypical views of them that would compromise the project. Numerous U.S. students feared that the Tunisians would dislike them by virtue of their being from the United States, fears that echoed the U.S. political analysis so prevalent at the time of the collaboration. Carl asserted that "most countries have negative views of the United States"; Matt wrote that "some countries hate Americans"; and Ned wrote that "Americans are often seen as arrogant, closed-minded, and short-sighted in other parts of the world. And, to some extent, that is true." Ron wrote, "I just hope none of them believe that political happenings surrounding our country are something that we have control over and don't look down on us because of our governments' [sic] decisions."

The U.S. students also were concerned that the ethos of Tunisian culture would interfere with communication. Although Phillip expressed concern that the Tunisian students would "be as lazy as we are," he concluded by suggesting that Tunisians would not be good-faith partners insofar as they would "not necessarily put forth an honest effort, *like we will*" (emphasis ours). Indeed, the potential lack of honesty among Tunisian students was mentioned frequently. This fact may reflect an assumption that economically developing countries are so positioned because their citizens are lazy, apathetic, or corrupt. Upon reading these data, the Tunisian co-author of this chapter recalled having encountered the same attitude during travel and stay in other comparatively affluent nations—not only in the United

States. Given the U.S. students' international communication case-study preparation, which focused on the pragmatic issue of saving face, it is not surprising that a few believed the Tunisians would shy away from giving helpful feedback in, again, an effort not to offend (see Table 4 for a breakdown of the fears voiced by the U.S. students and their possible sources). On the other hand, although Carl believed that, due to a fear of offending, neither side would be direct or necessarily honest with the other, he also believed that "they will have faith in us, just like we will in them."

A few students were concerned that religious differences might make communication difficult. Danny noted that "the main religion in Tunisia is Muslim [sic] which produces a fairly conservative culture which is a little different than here in America." Josey observed, "Since Tunisia is predominantly a Muslim culture, there are bound to be differences in their beliefs as compared to ours." Like a fear of causing offense, fear of culture-based conflict can inhibit communication by leading to avoidance and to low levels of engagement. Robert O'Dowd (2003), who conducted his study with the involvement of Spanish students, considers fears of offending the other to be inhibitors that can lead to "missed" communication. Josey noted, however, that "the differences in cultures could be a contributing or detracting factor, it all depends on how willing we are to work with each other and be considerate of one another."

Table 4. Possible Assumptions Behind U.S. Students' Reservations about the Collaboration

Fears	Possible causes
Fear of offending	Differences in language and culture lead to miscommunication.
Fear of being resented (hated) by Tunisians	Others perceive the U.S. role in the conflict in the Middle East and around the world as negative. Others dislike Americans because they believe all Americans are rich and arrogant.
Fear of differences in ethos	Others are potentially dishonest and lazy since they come from a country that is not as successful economically as the United States.
Fear that the Tunisians will not have adequate language skills to communicate in English	ESL speakers use a substandard variety of English and cannot be linguistically accomplished enough for equal participation with native speakers.
Fears that technology will fail	Technology in the economically developing world does not work.

LEARNING TO NAVIGATE DIGITAL DIVERGENCE

The U.S. students had serious concerns that technology would fail them in the end. Aspects of technology that students were especially dubious about were the effects of technological divergences on response patterns and the possibility that technologies would be incompatible. "I would say that my main fear," stated Josey, "is that due to technology differences, such as Web site viewing or even basic e-mail use and opening of attachments that this project won't work at all because we won't be able to communicate with the class in Tunisia." Danny was somewhat concerned, as well, observing that "the technology difference could make viewing of certain Web pages difficult for [the Tunisians] and even connection to the Internet could be a problem for them at times." Crystal, a computer-science major, had a more sanguine attitude regarding the technological aspects of the project. "As for the problem with technology," she mused, "[I] don't think that is really too big of a deal either. It might cause some delay in transfer of documents, but I think that we will ultimately be able to reach our final goal." The U.S. students observed that the Tunisians' technology would probably be good though on a more limited scale than that of the Americans; that the project would give the U.S. students the opportunity to learn about the realities of language and technology barriers between two very different countries; that e-mail communication was an especially challenging medium, even between speakers from the same cultural milieu; and that getting information from collaborators was going to require persistence—just as in the "real world." As Candy noted, "Communication is often filled with barriers, whether it is international communication or not." And Josey reflected that even the disadvantages of technological barriers would help students understand Tunisian "culture and what level of technology Tunisians have access to[,] which is important to know for a company that would be expanding there."

Not surprisingly, divergences in Internet access caused delays and frustration, for the low-tech computer lab limited the number of e-mails the Tunisians could exchange, and they often had to find alternative means of sending messages, as Darine explained:

> We had access to [the university computer] lab only two hours per week. . . . And in this lab, as you see, we have not enough PCs. We sometimes have to sit three per computer. [Besides], we are not really good at doing this [like the U.S. students are]. I don't know why I felt that—I felt that they were [more technically proficient] than us. Maybe they were better in this—technical writing. Sometimes their e-mails were a little bit late. . . . But I also think our e-mails, too, took a little time to send. Sometimes [the group I belonged to] had to make an appointment at the cybernet, or one of the houses where one of our colleagues have an Internet connection. But I do believe that [the U.S. students] have more equipment. This facilitates their task. And this is not the case here in Tunisia.

Ikbal, one of the few Tunisian students with an Internet connection at home, described her group's composing process in this way: "I write to [the U.S. students] from my house, [using a handwritten draft] the other members of the group write for me . . . but the last time [the U.S. students] did not respond." This was one instance of communication breakdown that required our intervention, as we will explain.

In practice, the computers available in the Tunisian lab often failed to work, and the Internet connection frequently was interrupted midclass. Difficulties with technology were exacerbated by a lack of technical support during the sessions and by the long-suspended purchase of software at that time in the Tunisian institution as a whole. The Tunisian students were well aware of this aspect of technological life at their school. As Darine pointed out, divergent digital environments exist even within the various colleges of the university she attended, which are located many kilometers from each other and among which students do not normally circulate. In her city, stated Darine,

> We have a School of Commerce, and the students there have more [computer] equipment than us. And they have sophisticated other things [media], I think. I want you to know that this is the case here only (addressing Richards). . . but this is not the case elsewhere in Tunisia. [Students in commerce] are specialized in such things. We are in literary (frowns) writing and everything to do with novel [literature] (rolls eyes).

Because of the comparatively strict disciplinary culture of the Tunisian university system, Tunisian students did not have the advantage, as many humanities students did in the U.S. classroom, to receive assistance from peers who were majoring in computer sciences. On their campus, the Tunisian business English students had access to computers only during the time of their lab; only a minority had a dial-up connection at home; and none had DSL. In contrast, in the U.S. classroom, each student had access to a computer with high-speed Internet. Almost certainly, all had access to comparable technology in their homes or dorms as well although personal computers for students were not as prevalent in the United States at that time as they have become since. Not once did technology itself fail in a significant way in the U.S. classroom.

The Tunisian students attempted to circumvent the problem of limited access, trying as best they could to participate "normally," resorting to wired homes of group members, as Ikbal reported, or cybercafés when necessary and possible. Yet cybercafé administrators who objected to the presence of more than one individual at a computer would often turn the students away. The fact that they were a group of three or four also complicated the work of the Tunisian students because they had to meet to discuss the U.S. students' e-mails, then explore the Web sites at a location where they could take notes, and, if this was a cybercafé, exit to draft the e-mail and return to send it. Usually, the students appointed the most computer-literate group member to do the sending.

When left to manage their own communication, the Tunisian students willingly and actively took on the role of independent learners. But their teacher, who had hoped they would negotiate their own learning, felt compelled to intervene at times by forwarding student e-mails to the U.S. students, a strategy that ultimately salvaged the project. In doing so, however, the Tunisian teacher found herself in the uncomfortable position of breaching her students' privacy even though they had requested her assistance, as she naturally read through the correspondence in the process of assisting with content or was privy to the information when transferring files. And as a result of helping students to correspond during class time, the teacher found that it was usually impossible to observe the potential of collaborative work or to record individual growth opportunities within groups, worthwhile scholarly activities discussed in the literature on computer-mediated communication and online learning (e.g., Kitade, 2008; O'Smith, 2005).

The Tunisian teacher, like her students who needed to purchase access to the Internet each time they sent a message from a cybercafé or from home, often purchased per-minute dial-up access at a substantial cost from her own home in order to forward her students' messages. As it happened, then, additional immediate financial burdens were situated rather squarely on the shoulders of the Tunisian participants. The U.S. teacher explained this situation to her students, a decision that prompted at least one to think more deeply about the power dynamics of intercultural digital communication. "If we are going to attempt to complete a project with groups of people who are paying for Internet on a per-minute basis at a location that may be some distance from their place of residence and who may not be comfortable with the e-mail process," Randy reflected, "We have to be willing to accept some delays and inconvenience."

The U.S. students also found the divergent digital environments they navigated with their international peers to be challenging. Andy, whose group seemed to have had the most successful relationship with their Tunisian collaborators, characterized the relationship midway through the semester in this way:

> Generally, after waiting for a reply from our Tunisian colleagues for a week or two, we email [their teacher], copying our message that we sent to her students. When we emailed [her], however, we received a reply within a few days, with our Tunisian collaborators [sic] replies. Since [her] reply is so speedy and includes her students' response, I believe that her students are responding to our emails promptly, but that their replies to us are not delivered.

This reasonably successful ad hoc process was the exception rather than the rule. Many U.S. students indicated frustration with the sending of attachments and the often extended delay between responses. "My group," stated Danny, "did get a response from them, but it was a blank e-mail with a Word document attached. However, the Word document didn't contain any data." Candy concurred, noting, "Our group sent them our first e-mail to begin the collaboration, but then the response we received was just their names. There was an attachment to the e-mail, but we were unable to open it." And, as Fred indicated, assignment deadlines were on the U.S. students' minds, so the wait for e-mail responses was at times unnerving.

Despite their frustrations, the U.S. students seemed not to blame the Tunisian students. Rather, they cited error or a lack of energy on their own side, an imbalance of power in terms of their advanced knowledge of computer technology, the likely sluggishness of Tunisian technology, unequal access, and the drawbacks of online communication itself as the probable culprits behind the collaboration's difficulties. "Our group has kind of been put on hold lately," observed Carl. "I think we put forth an effort to get the ball rolling, which appears to be more than they have done. But maybe we need to do a little more to encourage a response from them." He then considered that his group's e-mails may not have been written clearly:

> While thinking about this, I realized that it is possible our partners do not know how to reach us. I don't think I included an e-mail address in the letter, and maybe while the message was being sent from myself to [our teacher], to their teacher, to them, our e-mail address could have been lost.

Other U.S. students recognized that the playing field was not level and observed that they were probably at a significant advantage in terms of their technological know-how and access. For instance, Fred believed that the U.S. students were unfairly advantaged because "we have had a lot of experience with [e-mail] here at [state university] as e-mail has become one of the most popular means of communication." That so many of the U.S. students were computer-science majors was not lost on him, and he drew an unfavorable but affectionate comparison between the Tunisian business English students and his mother, who evidently was unable to grasp "simple things such as turning on a DVD player. . . . Some people," he mused, "may not be as technically inclined as others." Carl chose to disagree with Fred, however. "Like [Fred's] mom, my mom does not have the best understanding of technology," he wrote. "If I had to guess though, I doubt this is the problem with the Tunisians. It is more likely it is the result of different levels of technology."

The perceived inadequacy of Tunisian technology, and of the Tunisian students' limited access figured in the U.S. students' reluctance to blame their international partners. "Probably one of the biggest hindrances to this collaboration is the technological differences," stated Candy. Fred agreed, observing that both speed and functionality were key aspects of technology in the United States and thus, for instance, "if the media is not fast, it will not be effective" from the standpoint of U.S. users. The U.S. students cited likely access problems as additional contributors to the difficulties the collaborators were experiencing. "My group has not yet received a response from our Tunisian collaborators," wrote Chris. "I am under the impression that they do not have the access to the technology we so often take for granted." Matt stated, "Another thing they aren't able to do is use the Internet all the time. They aren't that familiar with it, so opening attachments could be a problem." Also giving his Tunisian peers the benefit of the doubt, Chris stated, "I think if we decided to use communication devices other than e-mail, we would have a more prompt response to our next attempt." Andy labeled the medium itself unreliable, and Bart noted that e-mail was an especially difficult means of correspondence, again potentially deflecting criticism of his Tunisian peers: "With electronic communication," he wrote, "not only do the people that are reading the document need to understand what is being sent, but the computer that is receiving the documents needs to be able to read the formatting of the document or the person wont [sic] have a chance."

Interviews with the Tunisian students revealed, however, that the delays often arose because students were not sure how to respond usefully to requests for expert content knowledge. Wafa discussed how the Tunisian students experienced delays because they were challenged by the "unusual" business-oriented content of the Web sites they were meant to explore. As Ikbal explained, "It is the first time we study e-commerce. We had no experience about this knowledge [perusing Web sites]. That's why we don't e-mail on time." Evidencing a comparatively communal orientation, she added that working in groups slowed the process further: "I wanted to respond for them, I shouldn't be uh [the one responding]. . . . I should wait for my colleagues. . . . They [my team members] feel there is ambiguity in the questions. . . . I wanted to respond very quickly, but what should I do?"

At least one Tunisian student felt that the questions they received from their U.S. peers were "too general" and wondered at times what they were being asked: "Sometimes," admitted Wafa, "I don't have an idea about what they were asking about. . . . I'd try to understand and to find an idea and, uh, by, uh, uh, looking at Web sites and it, uh, will finally—I can finally develop a general idea and reply to their question." In reference to the technologically pointed questions coming from the U.S. technical communication students, she elaborated, "The most difficult thing we experienced [was that] there was some Web site that you spoke [about], but we, as Tunisian students, don't know it very well. It was about the petroleum or something. . . . I remember that the questions . . . were not very clear. So we wondered what they wanted."

Danny felt that it was his responsibility and the responsibility of the other U.S. students to help Tunisian students feel positive about their roles in the collaboration. Regarding the issue of unusable attachments, he mused, "I think it would make the most sense to tell them that we didn't receive what they sent, even though we did receive it. If we told them that all we got was a blank file, they might be embarrassed for not being able to send the file correctly the first time. I think we can save face with them if we stretch the truth a little." His use of the phrase *we can save face with them* suggests that he saw the success of the collaboration as dependent upon not making his messages negative or critical of the Tunisian students; for instance, when it came to their perceived failure to open attachments. Chris made this imaginative leap: "I assume that our collaborators are just as frustrated as we are, and I am sure they are doing their best to overcome the difficulties we are having with our communication."

Indeed, the students often reminded each other to practice patience in the collaboration. "When we send our next email (which may be sent before we hear a reply)," wrote Randy, "We will be patient and keep it simple. . . . In some ways patience is something that needs to be inherent in this project. Also, I would say that understanding technological differences is the issue that would necessitate patience, and I would not consider it a separate issue." John stated, "I feel that there are many important qualities needed when engaging in a[n] international collaboration through electronic media. The first quality needed is patience. It is important to understand that it will take time for both parties to get on the same page in terms of a given collaboration." Matt observed wisely that "A relationship with someone overseas will not happen overnight. The need for patience will greatly reduce the stress involved in creating the relationship. The patience will also create a sense of normalcy because of not rushing into things and keeping the flow going in the right direction." Reflecting on the best strategy for prompting the Tunisian students to reply to his group's e-mail, Danny wrote,

> If we sent another email we would need to word it in a way that we don't come off as being pushy. This is because if they are frustrated with their technology the last thing they need is for us to badger them. . . . I also agree with Janine when she says that we will learn much if we are able to overcome these barriers between our two cultures/technologies.

John summarized the firsthand knowledge that the U.S. students were gaining about digital divergence:

> Here in the US our technology exceeds many other countries. . . . [An] important quality needed for international collaboration is to understand the technological difference between the two collaborating parties. As evident in our collaboration, technology differences between different countries can be very drastic.

It seems that many of the U.S. students had begun to recognize and to critique their positions as privileged users of digital technology. At the same time, the Tunisian students were gaining practice in developing creative ways of coping with the limits of digital technology in their own country. The U.S. students seemed to have grasped the concept of face saving and were actively involved in shoring up the self-confidence of their international partners. And among students on both sides of the collaboration, digital divergence at times provided a welcome alternative explanation for certain difficulties arising from the low linguistic or rhetorical competencies of individual Tunisian or U.S. students or their groups.

TAKING STOCK

This chapter has detailed our students' expectations of their South-North online collaboration as well as their interactions with digital divergence. Since the collaboration's conclusion, we have engaged in an extended process of reflection that has resulted in a set of recommendations that we hope will be useful to those who wish to engage in collaborations across two materially and technologically very different contexts. Many of the recommendations identified in Table 5 will also be appropriate for North-North collaborations; we consider them essential topics of consideration, however, in collaborations involving classrooms in postcolonial settings. For insights into the implementation of North-North collaborations, most of which will be relevant to South-North collaborations, see, for example, Paretti, McNair, and Holloway-Attaway (2007) Starke-Meyerring, Duin, and Palvetzian (2007), and St.Amant (2007).

Although assigned to students of practically oriented communications courses, the online collaboration we have described in this chapter had a self-consciously idealistic tone, for we both consider intercultural learning a key to successfully inhabiting a globalizing world and believe that all students, whether from the technologically accomplished North or the technologically enthusiastic South, should have the opportunity to become writers in the digital age. Chapelle (2003) refers to a "third set of abilities" (p. 16) separate from traditional communication skills and predicts that "participation" in the digital era will become a key pedagogical concern because it will drive learners' legitimate desires to develop an essential communicative competence. It must not be forgotten that the digital revolution and the globalization of world economies are mutually reinforcing (Crenshaw & Robison, 2006). Considering the circumstances of the economically marginal in both our countries, we conclude that, even in light of the agonizing process of globalization, economic isolation or maintenance of the status quo (were either possible) would in the long view be less desirable than strategic cultural transformation. Thus we advocate networked, informed, and critical citizenship for all who seek it.

A key goal for this chapter has been to explore the pedagogical benefits of South-North online collaborations. As we have mentioned, a limitation of North-North collaborations is that students and teachers may miss an opportunity to learn about technological and material landscapes that are very different from their own. Yet another reason for regret is a lost opportunity to come in contact with new ways of understanding the world. The collaboration we discuss, between students in the United States (where 0.6% of citizens are Muslim) and Tunisia (where 98% are Muslim) helped open our students' eyes to paradigmatic differences in worldviews, academic interests, social habits, educational systems, and workplaces, among other issues.

The clear majority of students from both classrooms believed that their intercultural online collaboration had been a positive experience. Matt, who earlier had observed that "a relationship with someone overseas will not happen overnight," summarized in his final portfolio what he had taken away from the collaboration: "good communication skills [are] essential in today's society to create a good relationship for the business world." According to him, the partners in such relationships

> need to relate to one another and be able to understand each other's ideas and concepts. Doing business with someone from another country can be difficult at times because of cultural differences. But keeping the ideas clear and having knowledge of the culture will better both partners in the relationship.

Like a few of their U.S. peers, the seemingly more affectively oriented Tunisian students spoke warmly of the chance to participate in a digital dialogue across national borders: "We were very, very happy to know that we will exchange ideas,

Table 5. Essential Topics for Consideration in Planning South-North Collaborations

Topic	Recommendation
Material conditions	For a South-North international online collaboration to begin to achieve the goals of critical pedagogy, students must develop an awareness of the material conditions prevailing in the lives of their partners. Students can study life in the partner country, exploring topics on which the U.S. students in our collaboration delivered in-depth oral presentations; e.g., the economy, business culture, political system, family life, religion, art and music, science and industry, and education. Since the time of our collaboration, digital technology has evolved to offer powerful new ways for students to gather such information. Now that satellite television is omnipresent and YouTube is accessible to anyone with a high-speed Internet connection, television shows (and advertisements, which can be especially illuminating) from other countries can also be incorporated into classroom discussions. Online discussions regarding key topics can be facilitated on blogs or Facebook or through other forms of social media.
Assessment systems	To avoid misunderstandings that arise from uninformed assumptions about other educational system, we suggest giving priority to grounding work that helps students understand each others' academic situations and objectives before engaging in cross-cultural communication. Students should be helped to recognize that educational systems can take very different approaches to assessment; differences that are likely to influence attitudes toward learning. For instance, because assessment in the U.S. system was ongoing, U.S. students in our collaboration were likely to be concerned that they might not be able to complete assignments on time and that they would receive a poor final grade as a result. Tunisian students, on the other hand, expected to receive only one grade: the outcome of their performance on a final, cumulative exam. For these students to participate actively in the e-collaboration, they needed to consider the project intrinsically valuable. Teachers of students in exam-focused systems might consider adding a bonus to the final grades of students who participate in online collaborations.
Worldview differences	Joint assignments should take into consideration any religious holidays on either side. A master calendar listing all such dates, national holidays, and important due dates for both classes can offset much confusion, as can the posting of all classroom assignments and syllabi on both sides of the collaboration. In fact, teachers can capitalize on this by proposing tasks that help students learn about each others' festivals as part of the grounding

Table 5. (Cont'd.)

Topic	Recommendation
	work mentioned above. Dates for school breaks and dates for the beginning and the end of the semesters and collaboration should be noted as well. A joint calendar will impress on students that all schedules in the world do not follow their own, and that time is not apportioned or understood in any universal sense—an important insight for intercultural communicators.
Friendships	The online collaboration should be understood primarily as a serious academic activity requiring students' attention and commitment. If there is a window of opportunity for sustaining relationships, however, students should be encouraged to feel free to correspond or to communicate in any way they see fit. Because teachers cannot restrict students' relationships over the Internet beyond the discipline imposed by the class project, students on both sides should be equipped with ways to decline unsolicited e-mail on both sides as part of their engagement with the pragmatics of cross-cultural e-mail communication. If teachers continue to collaborate over multiple semesters, a log with the names and perhaps personal or professional information (depending on participants' individual preferences) of each semester's collaborators can be maintained and shared with students so that they have the potential to continue communicating across cultures, within an expanding network.
Conflict	Students' expectations regarding online collaborations will be colored by their prior readings. Although the U.S. students had read and discussed a number of handpicked case studies in intercultural communication early in the semester, these were probably too negative in tone. Since the purpose of the case studies they read was to explore solutions to thorny problems, articles that concerned more affirmational (exemplars of successful) cross-cultural communication also should have been included in the curriculum so that students would not be led to believe that intercultural professional communication was basically a battleground. It is important that students collaborating across distances develop confidence and familiarity with others so that ultimately they can put the saving of face on an equal footing with the ability to question decisions and to offer alternatives. Palloff and Pratt (2005) remind us that the ability to navigate "disagreement or conflict about process or task" is crucial in online learning, for "the development of a 'groupthink' mentality, where everyone agrees in order to avoid conflict, often results in many members being dissatisfied with the results" (p. 15).

even through the Net only," stated Darine. Wafa concluded that "it was a beautiful experience [although] the connection between computers was not always good." And Siwar insisted, "I wasn't worried at all. I liked it very much. I liked this experience. I wanted it to be something beautiful between Tunisian students and American students."

By dealing more or less exclusively with others who are more or less like oneself, it is impossible to grasp the core realities of globalization, under which countries and individuals of vastly different means and mentalities work and struggle. In South-North online collaborations, the realities of globalization have a greater potential to surface. The ever-widening and often destructive path of globalization continually regenerates our patience for creating a blueprint for South-North collaborations in professional communication.

REFERENCES

Allen, I. E., & Seaman, J. (2008). *Staying the course: Online education in the United States, 2008*. Needham, MA: The Sloan Consortium.

Belz, J. A. (2003). Linguistic perspectives on the development of intercultural competence in telecollaboration. *Language Learning & Technology, 6*(1), 68–117.

Bosley, D. S. (2000). *Global contexts: Case studies in international technical communication.* Boston, MA: Allyn and Bacon.

Burnett, R. (2000). *Technical communication* (5th ed.). Boston, MA: Heinle.

Byram, M. (1997). *Teaching and assessing intercultural communicative competence.* Clevedon, UK: Multilingual Matters.

Byram, M., & Fleming, M. (Eds.). (1998). *Language learning in intercultural perspective.* Cambridge, UK: Cambridge University Press.

Cameron, D. (2002). Globalization and the teaching of "communication skills." In D. Block & D. Cameron (Eds.), *Globalization and language teaching* (pp. 67–82). New York: Routledge.

Chapelle, C. A. (2003). *English language learning and technology*. Amsterdam: John Benjamins.

Cope, B., & Kalantzis, M. (Eds.). (2000). *Multiliteracies: Literacy learning and the design of social futures*. London: Routledge.

Crenshaw, E. M., & Robison, K. K. (2006). Globalization and the digital divide: The roles of structural conduciveness and global connection in Internet diffusion. *Social Science Quarterly, 87*(1), 190–207.

Crystal, D. (1997). *English as a global language.* Cambridge, UK: Cambridge University Press.

Dautermann, J. (2005). Teaching business and technical writing in China: Confronting assumptions and practices at home and abroad. *Technical Communication Quarterly, 14*(2), 141–159.

Derbel, F., & Richards, A. R. (2007). Infusing a post-colonial component into English language teacher education curricula for a global century. *Radical Pedagogy, 9*(1). Retrieved September 8, 2008, from http://radicalpedagogy.icaap.org/content/issue9_1/derbel_richards.html

Fedderhöldt, K. (2001). An e-mail exchange project between non-native speakers of English. *ELT Journal, 55*(3), 273–280.

Fürstenberg, G., Level, S., English, K., & Maillet, K. (2001). Giving a virtual voice to the silent language of culture: The *Cultura* Project. *Language Learning & Technology, 5*(1), 55–102.

Giddens, A. (2000). *Runaway world: How globalization is reshaping our lives.* London: Routledge.

Hunsinger, R. P. (2006). Culture and cultural identity in intercultural technical communication. *Technical Communication Quarterly, 15*(1), 31–48.

Jackson, T. A. (2004). Imagining future(s): Toward a critical pedagogy for emerging technologies. In J. A. Inman, C. Reed, & P. Sands (Eds.), *Electronic collaboration in the humanities: Issues and options* (pp. 285–309). Mahwah, NJ: Lawrence Erlbaum.

Kaufman, L. (1998, March 17–21). *Email keypals in zone of proximal development.* Paper presented at the 32nd annual meeting of the Teachers of English to Speakers of Other Languages, Seattle, WA.

Kern, R. (1995). Restructuring classroom interaction with networked computers: Effects on quality and quantity of language production. *Modern Language Journal, 79*(4), 457–476.

Kern, R., Ware, P., & Warschauer, M. (2004). Crossing frontiers: New directions in online pedagogy and research. *Annual Review of Applied Linguistics, 24*, 243–260.

Kitade, K. (2008). The role of offline metalanguage talk in asynchronous computer-mediated communication. *Language Learning & Technology, 12*(1), 64–84.

Knowles, N., & Hennequin, M. W. (2004). New technology, newer teachers: Computer resources and collaboration in literature and composition. In J. A. Inman, C. Reed, & P. Sands (Eds.), *Electronic collaboration in the humanities: Issues and options* (pp. 91–110). Mahwah, NJ: Lawrence Erlbaum.

Kramsch, C. (1993). *Context and culture in language teaching.* Oxford, UK: Oxford University Press.

Kramsch, C., & Thorne, S. L. (2002). Foreign language learning as global communicative practice. In D. Block & D. Cameron (Eds.), *Globalization and language teaching* (pp. 83–100). London: Routledge.

Lee, L. (2004). Learners' perspectives on networked collaborative interaction with native speakers of Spanish in the U.S. *Language Learning & Technology, 8*(1), 83–100.

Lee, L. (2006). A study of native and nonnative speakers' feedback and responses in Spanish-American networked collaborative interaction. In J. Belz & S. Thorne (Eds.), *Internet-mediated intercultural foreign language education* (pp. 147–176). Boston, MA: Heinle & Heinle.

Liaw, M. L. (2006). E-learning and the development of intercultural competence. *Language Learning & Technology, 10*(3), 49–64.

McClanahan, L. G. (2001, June 25–27). *E-Pals: Examining a cross-cultural writing/literature project.* Proceedings of the 22nd NECC 2001: National Educational Computing Conference , Chicago. (ERIC Document Reproduction Service No. ED 462 938)

O'Dowd, R. (2003). Understanding the "other side": Intercultural learning in a Spanish-English e-mail exchange. *Language Learning & Technology, 7*(2), 118–144.

O'Smith, R. (2005). Working with difference in online collaborative groups. *Adult Education Quarterly, 55*(3), 182–199.

Palloff, R. M., & Pratt, K. (2005). *Collaborating online: Learning together in community.* San Francisco, CA: John Wiley & Sons.

Paretti, M. C., McNair, L. D., & Holloway-Attaway, L. (2007). Teaching technical communication in an era of distributed work: A case study of collaboration between U.S. and Swedish students. *Technical Communication Quarterly, 16*(3), 327–352.

Rassool, N. (1999). *Literacy for sustainable development in the age of information*. Clevedon, UK: Multilingual Matters.

Rickly, R. (2004). Computer-mediated communication as reflective rhetoric-in-action: Dialogic interaction, technology, and cross-curricular thinking. In J. A. Inman, C. Reed, & P. Sands (Eds.), *Electronic collaboration in the humanities: Issues and options* (pp. 35–48). Mahwah, NJ: Lawrence Erlbaum.

Scott, J. B., & Longo, B. (2006). Guest editor's introduction: Making the cultural turn. *Technical Communication Quarterly, 15*(1), 3–7.

Smith, B. (2008). Methodological hurdles in capturing CMC data: The case of the missing self-repair. *Language Learning & Technology, 12*(1), 85–103.

St.Amant, K. (2007). Online education in an age of globalization: Foundational perspectives and practices for technical communication instructors and trainers. *Technical Communication Quarterly, 16*(1), 13–30.

Starke-Meyerring, D., Duin, A. H., & Palvetzian, T. (2007). Global partnerships: Positioning technical communication programs in the context of globalization. *Technical Communication Quarterly, 16*(2), 139–174.

UNESCO. (2005). *Information and communication technologies in schools: A handbook for teachers or how ICT can create new, open learning environments*. Paris: Division of Higher Education.

U.S. Central Intelligence Agency. (2009b). The world factbook: Tunisia. Retrieved June 12, 2009, from http://www.cia.gov/library/publications/the-world-factbook/geas/Ts.html

Wang, Y. (2006). Negotiation of meaning in desktop videoconferencing-supported distance language learning. *ReCALL, 18*(1), 122–146.

Warschauer, M., & Kern, R. (2000). *Network-based language learning: Theory and practice.* Cambridge, UK: Cambridge University Press.

Yowell, B. (1995). The world at our fingertips. (ERIC Document Reproduction Service No. ED 378 957)

http://dx.doi.org/10.2190/CWDC8

CHAPTER 8

Meeting Online Friends Offline: A Comparison of South Korean and U.S. College Students' Differences in Self-Construal and Computer-Mediated Communication Preferences

Heeman Kim and William Faux

OVERVIEW

This chapter investigates the link between cultural and individual differences in relation to both the preferences of computer-mediated communication users who are accessing such utilities as Web-based chat services, instant messaging, e-mail, and Internet bulletin boards, and the creation of extended personal relationships in online and offline contexts. A survey was conducted to explore how members of different cultures use the Internet while interacting online. Respondents included 106 students from a large eastern university in the United States and 111 students from a midsize university in South Korea. Data indicated that members of an *interdependent* self-construal group were more likely than members of an *independent* self-construal group to post opinions on an Internet bulletin board. Moreover, interdependent group members were more likely than independent group members to have friends who they met online, and their online intimacy was more likely to extend to offline meetings. Results shed light on online activity that has been mostly ignored by cross-cultural communication researchers but embraced by practitioners.

With every newly adopted technology, innovative ways emerge in which individuals enact cultural practices, a fact suggesting that cultural variances in how a technology is utilized by diverse groups are to be expected. Often, however, cultural variances are examined through exclusively Western perspectives (Kim, 1999) and researchers run the risk of ethnocentrism (Ting-Toomey, 1997). For example, one widespread and often inaccurate perception among Westerners is that non-Western communicators talk less and are more passive than they themselves are (Burgoon, 1976; McCroskey & Richmond, 1987). Such generalizations tend to overlook contextual influences such as a culture's deference to authority or tolerance for ambiguity (Hofstede, 1980; Kim, 1999). Extending certain of these generalizations to non-Western technology-mediated situations such as online communication contexts is extremely problematic, especially given the growing number of intercultural interactions taking place online. As the number of World Wide Web users grows past 1.5 billion (Internet World Stats, 2009), there is a need to explore both how non-Western cultures are using computer-mediated communication utilities to build relationships and communicate online and how cultural variances influence that communication.

The growth of the World Wide Web has been attributed in part to the fostering of increased participation in social and political networks including both online and offline activities (Global Internet Statistics, 2004; Rheingold, 2000). For example, during the 2002 South Korean presidential election, Korean Internet users aggressively constructed their political and social agendas by voicing opinions online. At the same time, they offered offline support by gathering in a variety of physical locations after deciding to do so online (Lee, Lancendorfer, & Lee, 2005).

Such offline gatherings originating through the Internet (as opposed to mechanisms such as Twitter), in fact are a remarkably salient social phenomenon in South Korea among Internet users, who are in the process of constructing new social relationships (Jeong, 2000; Kim, 2001; Shin, 1999). These meetings' activities differ greatly depending on the features of the originating online community and may include going to the movies, having dinner, shopping, traveling, chatting, studying, and being physically intimate. An offline meeting of this type is known as *beongae* (or flash). This term implies an unexpected and immediate offline meeting of online participants at a specific physical location at a specific predetermined time. *Beongae* emerges as soon as one of the participants in either an asynchronous (e-mail, Internet bulletin board) or a synchronous (Web-based chatting, instant messaging) computer-mediated communication suggests an instant meeting for any reason and interested members immediately establish a place and time for *beongae*. Online suggestions to create a *beongae* event are observed easily on Internet bulletin boards, chat rooms, and instant messaging. South Korea is not the only country experiencing this contemporary online political and social phenomenon. *Beongae* events (or flash mobs) do occur in the United States and have been reported in the U.S. news (Farley, 2005; Rapp, 2005).

This chapter investigates the influence of cultural factors on the use of such computer-mediated communication, or CMC. John December (1997) defines CMC as a "process of human communication via computers, involving people, situated in particular contexts, engaging in processes to shape media for a variety of purposes" (para. 4). This definition refers to one-to-one or one-to-many communication performed through the computer, including Web sites, e-mail, electronic mailing lists, instant messaging, Internet bulletin boards, and Web-based chatting. Regarding electronic media and culture, the media of communication and our ways of using them help determine the kind of civilization in which we live (McLuhan, 1964).

Arnold Pacey (1992) argues that technological systems restructure the social world as an object of control (p. 7). James Katz (1997) suggests that communication technology not only is a means by which humans communicate with each other but also alters cultural environments and ways of life (p. 70). Pacey (1992, p. 7) and Katz (1997, p. 70) concur that electronic communication technologies reshape and restructure social-cultural patterns. Meanwhile, comparatively few researchers have studied CMC from a cross-cultural perspective (e.g., Aoki, 1995; Kim & Papacharissi, 2003; Stewart, Shields, & Sen, 1998). The majority of contemporary cross-cultural studies are concerned with developing an understanding of how individuals from diverse cultural backgrounds manage communication barriers in face-to-face situations (e.g., Kim, 1999; Park & Levine, 1999; Singelis, 1994; Singelis & Brown, 1995).

This chapter illustrates the link between cultural and individual differences as they relate to the preferences of CMC users accessing utilities such as Web-based chatting services, instant messaging, e-mail, and Internet bulletin boards and the creation of extended relationships within online and offline contexts. In so doing, the chapter provides an understanding of both communicative and cultural characteristics regarding interpersonal relationships online. Distinct differences regarding relationship extension to online/offline contexts between South Korean and U.S. college students are illustrated to inform research in the fields of international and intercultural communication and CMC.

CULTURAL AND INDIVIDUAL VARIABLES

According to Geert Hofstede (1980), the conceptual framework of *cultural variability* may shed light on people's beliefs, attitudes, and behaviors in intercultural contexts by examining the variances between areas common to many cultures. For Hofstede, these variances are presented as dialectical tensions between individualism and collectivism, short- and high-power distances, masculinity and femininity, certainty and uncertainty, and long-term versus short-term perspectives toward life. An individualistic orientation, which is common to many Western countries, tends to value an individual's achievement over that of the group whereas a collectivist orientation, which is common to many Eastern countries, tends to value the group's achievement over that of individual members. Harry Triandis (1995) metaphorically describes characteristics of individualism and collectivism by describing them as "tight" and "loose," respectively (p. 52).

On the relational level of communicating, individualism has been suggested to offer "advantages for situations where the individual is dealing with large collectives, such as the state, or for situations where achievement within the world economy is a central concern" (p. 171). A perspective that values both individual achievement and unclear superior/subordinate relationships may lead to increased violation of scripts in individualistic frameworks. In contrast, Triandis posits that collectivism offers "advantages for those social relationships that include small groups, such as family and co-workers, where people are dealing with face-to-face situations and with people they are going to be interacting with for a long time"

(p. 171). That is, collectivism offers the social scripts needed to promote group harmony, provide relational power distances, or clear boundaries of superior/ subordinate relationships, and foster certainty concerning what to expect in a given situation (Hofstede, 1980). In this sense, collective cultures would require a high tolerance for dealing with uncertainty. It is crucial to recognize that at times, predominantly collectivistic cultures display individualistic characteristics and that, at times, predominantly individualistic cultures demonstrate collectivist characteristics (Hofstede, 1980). Through the *dimensions* of individualism and collectivism, culturally developed beliefs and attitudes that influence communication traits can be illustrated.

Of equal importance is how individuals perceive themselves in relation to other members of their culture. Such perceptual awareness is needed if speakers/authors are to communicate in culturally competent ways (Gudykunst et al., 1996; Kim et al., 1996; Singelis & Brown, 1995). Many studies have examined individuals' self-construal as they operate in cultural contexts (Eaton & Louw, 2000; Heydenfeldt, 2000; Park & Levine, 1999; Singelis & Brown, 1995; Somech, 2000). Theodore Singelis (1994) notes that interdependent self-construal occurs when individuals regard themselves as members of their culture or society. According to Hazel Markus and Shinobu Kitayama (1991), the interdependent self has several characteristics. First, interdependent individuals perceive themselves not as entities separate from the immediate social situation, but as connected with other culturally bound members. Second, even though they have their own personal attributes (opinions and judgments), their demonstration of interdependency with other group members is salient in interactions. The interdependent self can be described as "sociocentric, holistic, collective, allocentric, ensembled, constitutive, contextualist, connected, and relational" (p. 227). Independent self-construal occurs when individual actors recognize themselves as unique entities independent and distinct from their native culture and society (Singelis, 1994). The independent self arises out of the recognition that the individual is separate from society and can be described as "individualist, egocentric, separate, autonomous, idiocentric, and self-contained" (Markus & Kitayama, 1991, p. 226).

Even though individuals have both an interdependent self and an independent self, one of the two is likely to be more salient depending on culture and situational context (Singelis, 1994). Researchers have identified certain relations between self-construal and individualism-collectivism. Self-construal is an individual variable, and individualism-collectivism a cultural variable (p. 586). Interdependent self-construal is prevalent in collectivistic cultures whereas independent self-construal is prevalent in individualistic cultures. Singelis and William Brown (1995) find that collectivist cultures are positively and significantly associated with interdependent self-construal and negatively and significantly associated with independent self-construal (p. 379). In addition, Min Sun Kim (1999) finds that individuals who rate as interdependent are more concerned with building relationships with others.

Since *culture* refers to an ongoing process that amalgamates existing values, attitudes, beliefs, and symbols with newly introduced artifacts, research on the cultural uses and effects of new communication technologies is far from unprecedented.

When new communication technology is introduced, a body of literature shows that its use and content rely heavily on cultural characteristics (Cassirer, 1977; Worth & Adair, 1970). Current and emerging research on the Internet demonstrates a range of answers to questions regarding the cultural uses and effects of electronic media (Cho & Cheon, 2005; Choong & Salvendy, 1998; Kim & Papacharissi, 2003; Polyorat & Alden, 2005; Sun, 2001; Zhang & Shavitt, 2003).

In fact, a number of researchers have examined the cultural uses of the Internet. Heeman Kim and Zizi Papacharissi (2003), for example, studied how cultural background influences self-presentation constructed through personal home pages by college students. The authors found that comparatively ambiguous visual formats for self-presentation are indicators of collectivistic culture whereas direct and explicit textual formats are indicators of cultures tending toward the individualistic (p. 114). Chang-Haon Cho and Hongsik Cheon (2005) analyzed corporate Web sites created by members of collectivistic and individualistic cultures. They found that collectivistic Web sites tend to use consumer-to-consumer interactive features, such as online discussion, chat rooms, Internet bulletin boards, and product user groups in which users swap ideas and feedback with peer consumers. Their data suggest that members of collectivistic cultures are more likely than members of individualistic cultures to seek group conformity and to engage in consumer group activities (p. 111). These studies indicate that even though members of different cultures use the same technology, they construct online presences differently by using culturally influenced message designs reflecting their individualistic and collectivistic cultural characteristics (Cho & Cheon, 2005; Choong & Salvendy, 1998; Kim & Papacharissi, 2003; Polyorat & Alden, 2005; Sun, 2001; Zhang & Shavitt, 2003).

As has been illustrated, individualism/collectivism variables have been used effectively to illustrate broad cultural perspectives and differences associated with online communication. In the study chronicled in this chapter, we adopt self-construal theory to explain cultural differences and similarities between the personal communication preferences adopted when individuals are online. In this regard, the current study extends CMC research into both the cultural and individual levels by placing them in dialectical tension (Baxter, 1994). That is, descriptions illustrating how individuals construct their affiliation as group members and how they establish their individuality emerge through their usage of CMC utilities.

COMPUTER-MEDIATED COMMUNICATION AND USER PREFERENCES

Internet research has often shown that CMCs have functioned as powerful alternative channels of communication for specific groups and individuals by providing a voice, a sense of anonymity, and empowerment to those reluctant to engage in face-to-face communication (McKenna & Bargh, 1999; Papacharissi & Rubin, 2000; Valkenburg & Soeters, 2001; Walther, 1992; Weiser, 2000). Lisa Flaherty, Kevin Pearce, and Rebecca Rubin (1998) examine whether in fact the Internet can function as a rich alternative to face-to-face communication. Their research suggests that, depending on the motives and needs of communicators, CMC usage can diverge significantly. Such findings challenge the notion that different media can "serve the same need equally" (p. 251).

Additional studies report that users of the Internet often communicate with friends and family because access is faster, easier, and cheaper than other technology-mediated communication options such as the telephone and snail mail, especially so when geographical factors include crossing international boundaries (Papacharissi & Rubin, 2000; Stafford, Kline, & Dimmick, 1999). In this regard, CMC usage maintains family ties when one is domiciled in a foreign country.

The various forms of CMC can be categorized as asynchronous or synchronous. Asynchronous CMC forms include bulletin board systems, electronic mailing lists, and e-mail; synchronous forms include Web-based chatting and instant messaging. Asynchronous CMC allows users to control their time and space in sending and receiving messages, but users need to spend more time interacting than with synchronous CMC. In synchronous, real-time communications, users must be present and involved simultaneously.

A variety of features and functions, such as whether a given CMC is synchronous or asynchronous, influences preferences regarding CMC use. Synchronous CMCs allow interlocutors to state their opinions and respond to others immediately while also providing the option to leave an interaction without notice (Rintel & Pittam, 1997). For example, interactions in chat rooms allow users to argue their opinions with comparative freedom, potentially disregarding the opinion and even presence of other members. As a form of synchronous CMC, Web-based chat rooms are public spaces in which participants meet virtually to initiate and to maintain personal relationships, including romance and friendship. Because chat rooms are considered a public space, however, participants may feel less of a sense of belonging than in private spaces (Haythornthwaite, 2000). And due to a dearth of visual cues, participants in chat rooms can find it more difficult to initiate interactions than in face-to-face communication; as a result, such online conversations often end comparatively abruptly. To compensate for deficits in physical presence, current advances in asynchronous and synchronous technologies allow users to exchange self-portraits (e.g., GIF, JPEG files), host live video/voice chat and Webcams, and share streaming blogs.

The construal of individuals interacting online may influence how certain CMCs are adopted and used to fulfill cultural and personal needs and motives. As mentioned, Kim and Papacharissi (2003) found that Web authors in South Korea were less likely to utilize direct descriptions of their identity, heritage, and residence and were also less likely to present personal stories and information than their U.S. counterparts were. In contrast, U.S. authors were more likely to provide personal information and more likely to structure their online identity by providing intimate descriptions. These researchers argued that self-construal values influence the design of personal homepages (p. 114).

Graham Barwell and Kate Bowles (2000) posited that online communities traverse national and cultural boundaries. That is, users around the world log onto the online community, post their own ideas about a topic, and are at the same time exposed to diverse opinions. Accordingly, since Internet bulletin boards often serve specific interest groups, members often feel affinity toward group identity and community that may transcend geographical and national barriers (Baym, 2000; Ogan, 1993; Rheingold, 2000). Internet bulletin boards allow users to review messages posted by others and allow readers to upload their own opinions. According to Christine Ogan (1993), the content of Internet bulletin boards is more diversified and personalized than in other forms of mass communication but less so than in face-to-face exchanges. Other studies show riskier verbal expression in one-to-many CMCs than in one-to-one CMCs (Choi, Lee, & Lee, 2004; Thompsen & Ahn, 1992). For instance, in examining a Thai newsgroup, Soraj Hongladarom (1998) found that members openly and freely discussed forbidden social issues such as the government and the royal family. These topics are officially prohibited in offline discussions. Online discussion also accelerates the formation of both group and national identity through the use of members' native languages.

As we have just illustrated, CMC options are used in unique ways; moreover, individuals utilize the features of CMC options based on their cultural and social communication needs. Tapio Varis (1999) suggests that in fact unique individualistic and collectivistic identities can be constructed in the Internet environment. This emergent area of research suggests that people's preferences for using CMCs may correspond to either independent self-construal or interdependent self-construal. Our first research question addresses such a probe:

RQ1: Is there a difference in CMC preference depending on two categories of self-construal (individuals fitting the independent self-construal category and those fitting the interdependent self-construal category)?

While members of a given culture might prefer to use a specific CMC, these two categories of self-construal have different traits that maintain long-term and short-term interactions and intimacy. That is, of particular interest is the extent to which various forms of online communication will be significant for members of a culture situated within a specific context. Not only are CMCs used to initiate one-time interactions, but also, because of increasingly inexpensive and high-speed service, they are used to maintain interpersonal relationships (Rheingold, 2000; Stafford et al., 1999). In fact, some CMC users regard the Internet as a vehicle to link online intimacy to offline meeting, and vice versa. By exploring what CMC users are more likely to use to initiate offline meetings, this study examines the connection between CMC preference and offline meeting experience. This online/offline relational phenomenon is observed frequently in South Korea, as discussed earlier, and might be explained through cultural characteristics such as interdependent and independent self-construal.

CMC users' maintenance of online relationships in and through both task-oriented (time-limited) and socially-oriented (time-unlimited) interactions has been examined by Joseph Walther (1992, 1996). Frequent message exchanges using CMC encourage users to develop online rapport, albeit more slowly than offline (Walther, Anderson, & Park, 1994). Literature also focuses on Internet users' initiation and maintenance of personal networks in chat rooms (Rintel & Pittam, 1997), electronic mailing lists, and Internet bulletin boards (Ogan, 1993). In their research examining the costs and benefits of Web surfing, however, Elizabeth Perse and Douglas Ferguson (2000) demonstrate that the Internet is not an equitable and rich replacement for social contact. Therefore, exploring whether individuals, depending on their self-construal, still feel the need to meet people offline is warranted. In this regard, we advance our second research question:

RQ2: Is an offline meeting experience dependent on the two categories of self-construal?

To provide a cultural context for this study, a brief description of the Korean computer-mediated environment is outlined as well as the different usages of CMC in South Korea and the United States. According to a 2007 report from the National Internet Development Agency of Korea (NIDAK), 79.8% of Korean households were connected to the Internet. Among those connections, 84% used high-speed connectivity, including XDSL (82.9%) and cable (21.4%). Interestingly, Internet access takes place in homes, workplaces, locations like *PC Bang*, and Internet cafés more frequently in South Korea than in the United States. The widespread use of cellular phones and PDAs offers additional Internet access in South Korea and illustrates another difference between the way that CMCs are used in South Korea and in the United States. One similarity between the two environments, however, is that Internet users tend to begin their Web surfing with commercial portal services (Internet).

Most portal services are major vehicles for online communication through e-mail, Internet bulletin boards, instant messaging, chat rooms, or electronic mailing lists. U.S. companies such as Yahoo!, AOL, MSN, and Lycos have extended their online territory to Asian countries such as South Korea. Even though the portal services provide features like those provided in the United States, they customize categories and topics to accommodate the cultural values and preferences of Korean users, using indigenous graphics, Korean domestic news stories, and Korean holidays. Korea has, in addition to imported services, many domestic portal services such as Daum, Naver, Hanmir, Damoim, Netian, Paran, and Entica. Korean domestic portal services offer more customized features and better reflect Korean communication needs online. As of December 2007, Daum ranked first in frequency of visits followed by Naver (Internet). Like U.S. computing services, the Daum site allows users to log onto its electronic mailing lists, chat rooms, e-mail, news services, and diversified bulletin boards; this service is named Daum Café and has a positive reputation for its effectively customized categories and topics, which foster involved, active participation.

METHODS

In accomplishing this study's purpose of exploring CMC utilities and offline meeting experiences in relation to cultural factors, a survey was conducted to assess how members of different cultures use the Internet while interacting online.

Respondents

Because of limitations involving access, geographical boundaries, and time, this study used a sample of convenience: Namely U.S. and South Korean college students. The authors believe that such individuals are representative of the Internet population, however, because they are active in online interactions and often have easy and free Internet access in school (Perse & Ferguson, 2000). In total, 217 respondents from South Korea and the United States were surveyed. Respondents included 106 students (34 males, 68 females, 4 not responding) from a large eastern university in the United States and 111 students (48 males, 61 females, 2 not responding) from a midsize university in South Korea. The mean age of the Korean sample was 22.05 (SD = 3.84), and ages ranged from 19 to 31; the mean age of the U.S. sample was 22.82 (SD = 3.11), and ages ranged from 19 to 38. The 106 U.S. respondents' reported racial makeup was 68.2% white, 19.1% African American, 3.5% Hispanic, 3.8% Asian, and 0.7% other. The South Korean participants were racially homogeneous.

Procedure

The survey was administered both in Korea and the United States. Originally developed in English, it was translated into Korean by a bilingual third party. To avoid instrument errors, the lead author conducted a test with 22 Korean students residing in the United States to confirm the instrument's accuracy in translation and reliability. Their agreement on the instrument's accuracy of translation (98%) proved satisfactory.

The questionnaires were administered by the lead author in the United States. A paid research assistant administered the questionnaires in Korea. The research assistant was trained by the lead author to conduct the survey and was informed of the purpose of the study through phone conferences and e-mail. The Korean respondents were drawn from sophomore-level journalism classes and usually spent 1.3 hours a day on the Internet, mainly information seeking. Three undergraduate classes in Korea were selected with the instructors' permission. Respondents received extra credit for participating in the study. The U.S. respondents were volunteers from sophomore-level journalism classes and spent on average 1.7 hours a day on the Internet, for both information seeking and entertainment. Three undergraduate courses were selected for the distribution of questionnaires; extra credit was suggested, and the instructors made the respective final decisions concerning whether to grant it.

Participants were given the survey packet, which consisted of five pages. The research assistant in Korea and the lead author in the United States briefed the students by reading a statement describing the purpose of the study and indicating that participants had the right to withdraw their consent at any time for any reason. The questionnaires were handed out at the beginning of the class and collected about 20 minutes later. The lead author in the United States gathered 106 questionnaires, and the assistant in Korea gathered 111 questionnaires and sent them to the author by mail.

Measurements

In Section 1 of the questionnaire, open-ended questions were asked along with closed-ended questions, to ascertain respondents' experiences with the Internet. The questions elicited information such as how much time participants thought they had spent surfing the Internet the day before the study; what their main purposes were for surfing the Internet; and how frequently they were involved in Internet bulletin boards, electronic mailing lists, e-mail, instant messaging, and Web-based chatting. This section also included questions about the respondents' degree of dependence on the Internet as an information resource compared with newspapers, and their use of the Internet as an entertainment resource compared with television. Lastly, a question focused on respondents' experience with and frequency of offline meetings, such as whether or not respondents had ever met online friends offline and whether or not respondents had ever met offline friends online. Questions not requiring open-ended responses used a 7-point Likert scale ranging from 1 ("strongly disagree" or "never") to 7 ("strongly agree" or "most frequently").

The authors adopted Kim, Aune, Hunter, Kim, and Kim's (2001) measurement to assess self-construal. The personal reaction inventory originally developed consisted of 29 items; 7 items, however, did not apply to online interactions and were eliminated for the purpose of this study. As a result, Section 2 of the survey consisted of 22 questions measuring the degree of interdependence or independence of respondents in relation to online interactions.

Finally, Section 3 of the questionnaire probed demographic factors such as gender, age, academic status, income, and ethnicity.

Statistical Analysis

To test the research questions, a factor analysis was performed to indicate outlying variables. An analysis of variance (ANOVA) was then used to reveal the different CMC preferences and offline meeting experiences based on two types of self-construal.

RESULTS

As shown in Table 1, a principal component factor analysis with varimax rotation was used, consistent with Kim et al. (2001), to sort variables into distinctive sets in the personal reaction inventory. The first and second factors were composed of 15 items with a total reported variance of 46.85%. Table 1 shows self-construal with factor loadings. The first factor, labeled "interdependent self-construal," was

Table 1. Factor Analysis of Personal Reaction Inventory

Statements	Factor 1	Factor 2
Interdependent self-construal		
1. My relationships with those in my group are more important than my personal accomplishments during online interaction.	0.741	0.132
2. My happiness depends on the happiness of those in my group during online interaction.	0.841	9.843; E-02
3. I often consider how I can be helpful to specific others in my group in online interaction.	0.695	0.253
4. I am careful to maintain harmony in my online group.	0.800	0.107
5. When I am with my chat peers, I watch my words so I won't offend anyone.	0.742	–5.662; E-02
6. I would sacrifice my self-interests for the benefit of my group.	0.664	4.183; E-02
7. I try to meet the demands of my group, even if it means controlling my own desires in online interaction.	0.650	3.768; E-02
8. I think it is important for members of an online chat group to follow Web etiquette.	0.689	–0.248
9. The security of being an accepted member of a chat room is very important to me.	0.681	8.102; E-02
Independent self-construal		
1. I voice my opinions during online interaction.	–1.307; E-02	0.669
2. My personal identity, independent of others, is very important to me in online interaction.	0.175	0.712
3. I offer my opinion in any way I want in online interaction.	–0.242	0.685
4. I act as a unique person, separate from others in online interaction.	–6.198; E-02	0.746
5. A major goal for me in online discussion is to argue my opinion strongly.	7.122; E-02	0.653
6. I enjoy being admired for my unique opinion in online discussion.	7.617; E-02	0.820
Eigenvalue	5.932	3.714
Variance explained	**29.97%**	**16.88%**

the major factor, explaining 29.97% of variance. The mean score for the nine items' interdependent self-construal was 3.76 (SD = 1.50), with 7-point Likert scales ranging from 1 ("strongly disagree" or "never") to 7 ("strongly agree" or "most frequently"); coefficient alpha was 0.89. The second factor, named "independent self-construal," was composed of six items that explained how respondents thought of themselves in online interactions (M = 4.56, SD = 1.21), with 7-point Likert scales ranging from 1 ("strongly disagree" or "never") to 7 ("strongly agree" or "most frequently"). The independent self-construal had a 0.82 alpha value and accounted for 16.88% of reported variance. The third factor reported a variance of 8.34%; this was eliminated because it contributed a small reported variance and was not conceptually part of this study.

For the sake of consistency in statistical analysis, items representing independent self-construal were addressed. The first step was to recode the values of items representing interdependent self-construal so that a low score indicated interdependent self-construal. Next, the mean self-construal scores from all 15 items were summed, divided by 15 to create a self-construal index, and then z-transformed. Because z-values for the self-construal scale ranged from –3.43 to 2.98, z-values ranging from –3.43 through 0.0 were recoded as interdependent self-construal, and those ranging from 0.0 to 2.98 were recoded as independent self-construal. In all, 97 participants (44.7%) were rated as the independent self-construal group, and 120 participants (55.3%) were rated as the interdependent self-construal group. As supported by previous cross-cultural studies, respondents from the United States significantly and positively presented independent self-construal while Korean respondents (from a collectivistic culture) significantly and positively presented interdependent self-construal ($X^2[1, N = 217] = 8.71, p < 0.003$).

Survey data are presented and discussed according to the research questions presented earlier. Participants believed that they were good at searching for information on the Internet ($[M = 5.58, SD = 1.25], t[214] = 65.28, p < 0.000$) and stated that they depended more on the Internet as a news resource than on television ($[M = 3.49, SD = 1.91], t[212] = 26.66, p < 0.000$). The interdependent self-construal group ($M = 4.03$, $SD = 1.83$), however, was more likely to rely on the Internet as an entertainment resource than the independent self-construal group was ($[M = 2.78, SD = 1.81], F[1, 209] = 24.52, p < 0.000$). The interdependent self-construal group ($M = 3.26$, $SD = 1.91$), by contrast, was more likely than the independent self-construal group ($[M = 2.19, SD = 1.53], F[1, 213] = 19.74, p < 0.000$) to use the Internet for socialization.

RQ1 concerned whether there was a difference in CMC preference depending on the two categories of self-construal. Table 2 presents the means and standard deviations for CMC uses. In response to RQ1, an ANOVA was performed on CMC preferences. As shown in Table 3, the interdependent self-construal group was more likely to post its opinions on the Internet bulletin board than the independent self-construal group was ($F[1, 213] = 34.15, p < 0.000$). There was no significant difference in use of e-mail ($F[1, 211] = 1.63, p < 0.20$), use of Web-based chat rooms ($F[1, 213] = 3.08, p < 0.08$), or use of instant messaging ($F[1, 201] = 0.34, p < 0.56$).

Table 2. Mean and Standard Deviation for CMC Uses

	CMC	Mean	Std. Dev.
Independent self-construal	Internet bulletin board	1.81	1.24
	E-mail	3.36	2.15
	Chat room	2.73	1.83
	Instant messaging	2.63	1.29
Interdependent self-construal	Internet bulletin board	3.08	1.81
	E-mail	3.71	1.64
	Chat room	3.10	1.96
	Instant messaging	2.75	1.58

Table 3. ANOVA of CMC Uses in Self-Construal

	Degree of freedom	*F*-value	Sig.
Bulletin board	$Df_B = 1$, $Df_W = 213$	34.15	0.00
E-mail	$Df_B = 1$, $Df_W = 211$	1.63	0.20
Chat room	$Df_B = 1$, $Df_W = 213$	3.08	0.08
Instant messaging	$Df_B = 1$, $Df_W = 201$	0.34	0.56

RQ2 asked whether offline meeting experience depended on the two categories of self-construal. Members of the interdependent group ($M = 1.43$, $SD = 0.49$) were more likely to have friends that they met only online (people they had never met in person) than members of the independent group were ([$M = 1.21$, $SD = 0.41$], $F[1, 213] = 12.39$, $p < 0.001$). Also, members of the interdependent self-construal group ($M = 1.39$, $SD = 0.49$) were more likely to meet online friends offline than members of the independent self-construal group were ([$M = 1.18$, $SD = 0.39$], $F[1, 213] = 12.01$, $p < 0.001$); that is, their online intimacy was more likely to extend to offline meetings. For CMC preferences, an offline meeting experience was coded as a dichotomous variable (no/yes); the dependent variable included CMC preferences. Data indicated that respondents who had offline meetings were more likely to use the Internet bulletin board ([$M = 4.27$, $SD = 1.42$], $F[1, 213] = 12.01$, $p < 0.001$) and that respondents who had not experienced offline meetings tended to prefer e-mail ([$M = 3.73$, $SD = 1.27$], $F[1, 213] = 17.01$, $p < 0.001$). There was, however, no interaction suggesting a relation between either self-construal or offline meeting, and CMC use.

DISCUSSION

The central focus of this study was to identify whether cultural characteristics such as independent and interdependent self-construal are extended to online interactions and whether CMC use differs between independent and interdependent self-construal groups. The data illustrated that Korean respondents showed significant interdependent self-construal online; at the same time, U.S. respondents did not demonstrate independent self-construal online. Because culture is not static but changes over time, it is important to identify areas of change, and especially when new technologies are adopted. Equally important is identifying comparatively enduring cultural characteristics (Collier, 2000). An example of the latter for the Korean respondents is that of interdependence.

In general, previous studies have identified that a primary feature of the Internet is its use in information seeking; a secondary function is in entertainment and recreation (Papacharissi & Rubin, 2000; Perse & Ferguson, 2000; Stafford et al., 1999). This study found, however, that the interdependent self-construal group, whose members reflected a prominent cultural characteristic of South Korean society, used the Internet primarily as an entertainment resource. This finding can be explained by a recent report published from NIDAK, which notes that 41.3% of young Korean Internet users frequently gain access through Internet cafés, or *PC Bang* (Internet Statistics Information System, 2008). They mainly enter Internet games and chat rooms there because *PC Bang*s are known to be well-equipped and to host a variety of online games for a fee. In short, *PC Bang* users are more involved in entertainment and socially oriented tasks than in information seeking. The findings and explanation demonstrate the impact of contextual forces, such as an economic or business infrastructure, upon a culture's use of technology.

As described earlier, interdependent self-construal is characterized by "tightness," implying interconnectedness (Triandis, 1995). This study illustrated that individuals with high interdependent self-construal tended to use a variety of CMCs, especially Internet bulletin boards. Members of the interdependent self-construal group also were more likely to post their opinions on Internet bulletin boards than members of the independent self-construal group were. Because they are both group- and context-specific, Internet bulletin boards are a more useful tool for individuals possessing and developing group identity. That is, Internet bulletin boards are normally used by individuals sharing similar group interests (Barwell & Bowles, 2000; Varis, 1999). Caroline Haythornthwaite (2000) found that members of closely knit groups use "asynchronous CMCs in a near-synchronous manner" (p. 221). Members of the interdependent self-construal group frequently sent and quickly replied to messages, thereby shortening the temporal gap of Internet bulletin boards and reaffirming group affiliation. Individuals possessing a collectivistic communicative characteristic would have a sense of tight membership when interacting within an online community, and the cultural characteristic of interdependent self-construal would be reinforced through Internet bulletin boards.

The last major finding of this study illustrated that people with high interdependent self-construal were more likely to meet face-to-face with people they originally met online. Also, members of the interdependent group had more online-only friends (friendships with people whom they had never met face-to-face) than did members of the independent self-construal group. That is, the interdependent self-construal group members' online relationships were more likely to extend to offline meetings (at times involving intimacy). Thus the interdependent self-construal group was more competent in its use of CMCs to maintain personal relationships both online and offline than the independent self-construal group was.

Many cross-cultural studies have claimed that people with high interdependent self-construal are passive communicators. That is, they talk less, engage in less information seeking (Burgoon, 1976), and disclose less (McCroskey & Richmond, 1987) than those who have high independent self-construal. Critiquing a Western viewpoint and culturally biased scales that stress the *frequency* of speaking, Kim (1999) has argued, in contrast, that interdependent self-construal demonstrates a communication competence acknowledging the intention of one's communicative counterpart. This characteristic influences the generation of an optimal message for achieving communicative goals (p. 80). Kim's argument supports the findings of the current study. Significantly, it is problematic to claim that people who rate high in interdependent self-construal are passive communicators, for they are more likely to meet people (offline) in face-to-face contexts. Considering the results of this study, and the characteristics of both collectivistic cultures and interdependent self-construal, we conclude that individuals with interdependent self-construal may be better able to translate online-relationships into offline, face-to-face relationships than those with independent self-construal are. An additional motivation for interdependent self-construal respondents to meet offline is that, as part of a collectivistic culture, they are more reliant upon contextual cues obtained from rich communication environments. Since CMCs provide limited contextual cues, the interdependent respondents may try to meet their counterparts beyond the computer monitor to satisfy their contextual and communicative needs.

Interestingly, respondents who had offline meetings were more likely to use Internet bulletin boards, but respondents who had no experience with offline meeting tended to use e-mail. This can be explained partly in terms of the features of Internet bulletin boards and e-mail. E-mail is considered task-oriented and a one-to-one communication tool that allows users to maintain existing relationships (Haythornthwaite, 2000). Internet bulletin boards can be regarded as a one-to-mass communication tool to spread and voice a user's opinion or idea so that Internet users may easily discern different voices from their opinion and find group companions who share their opinions. In this regard, Internet bulletin board users can easily meet both supporters and adversaries online (Barwell & Bowles, 2000; Varis, 1999). Hence, bulletin board users can foster group membership while maintaining and sometimes strengthening collectivity offline.

CONCLUSION AND LIMITATIONS

The findings of this study partly support and strengthen previous findings about interdependent and independent self-construal (Eaton & Louw, 2000; Heydenfeldt,

2000; Park & Levine, 1999; Singelis & Brown, 1995; Somech, 2000). The most interesting and valuable finding of this research, however, is that interdependent self-construal respondents were more likely to use Internet bulletin boards to communicate with group members and to develop and maintain both online and offline relationships. Additionally, the interdependent self-construal respondents demonstrated collectivistic characteristics and were more reliant on contextual cues obtained from rich communication environments; therefore, they met their counterparts beyond the computer monitor to satisfy their communication needs. The differences between interdependent and independent self-construal were characterized through distinctive CMC preferences. This study illustrates the preferences and frequencies of utilizing the Internet bulletin board; future research might explore differences in content qualitatively.

As in any research project, certain limitations must be addressed. First, a larger sample is needed for future research in order to generalize the findings derived from this study. Such data could be obtained from a more diversified population such as children, the elderly, the tech-savvy, and so on. Obtaining data from multiple sources would generally help ensure less biased results.

Second, this study focused on temporal gaps within the usage of a variety of CMCs. Many more features of CMCs could be explored, including spatial differences. That is, individual behavior online may depend on the public or the private nature of space online and may be influenced by the physical location in which one is logging onto the Internet. Future research may consider exploring this spatial feature of CMC. Third, examining geographic factors to explain the phenomenon of offline meeting in collectivistic cultures is warranted. For example, Korea is geographically less dispersed than the United States. This feature may contribute to facilitating face-to-face interaction via easy and inexpensive transportation; consequently, such offline interaction may be more likely to take place in Korea. It would be interesting to compare CMC uses among Asian countries to find whether there are any intra-Asia differences. Lastly, other characteristics within Hofstede's (1980) theory of cultural variability, such as power distance, reactions to uncertainty, and gender, can be explored in relation to CMC preferences.

REFERENCES

Aoki, K. (1995). Synchronous multi-user textual communication in international tele-collaboration. *Electronic Journal of Communication, 5*(4). Retrieved November 12, 2007, from http://www.cios.org/www/ejc/v5n495.htm

Barwell, G., & Bowles, K. (2000). Border crossings: The Internet and the dislocation of citizenship. In D. Bell & B. Kennedy (Eds.), *The cybercultures reader* (pp. 702–711). London: Routledge.

Baxter, L. A. (1994). Thinking dialogically about communication in personal relationships. In R. Conville (Ed.), *Structure in human communication* (pp. 23–38). Westport, CT: Greenwood.

Baym, N. (2000). *Tune in, log on: Soaps, fandom, and online community*. Thousand Oaks, CA: Sage.

Burgoon, J. K. (1976). The unwillingness-to-communicate scale: Development and validation. *Communication Monographs, 43,* 60–69.

Cassirer, H. (1977). Radio as the people's medium. *Journal of Communication, 27*(2), 154–157.

Cho, C., & Cheon, H. (2005). Cross-cultural comparisons of interactivity on corporate Websites: The United States, the United Kingdom, Japan, and South Korea. *Journal of Advertising, 34*(2), 99–115.

Choi, J., Lee, C., & Lee, J. (2004). Influence of poll results on the advocates' political discourse: An application of functional analysis debates to online messages in the 2002 Korean presidential election. *Asian Journal of Communication, 14*(1), 95–110.

Choong, Y., & Savendy, G. (1998). Design of icons for use by Chinese in mainland China. *Interacting with Computers, 9*(4), 417–430.

Collier, M. J. (2000). Understanding cultural identities in intercultural communication: A ten-step inventory. In L. A. Samovar & R. E. Porter (Eds.), *Intercultural communication: A reader* (pp. 16–33). Belmont, CA: Wadsworth Publishing Company.

December, J. (1997, January). Notes on defining of computer-mediated communication. *CMC Magazine.* Retrieved November 2005, from http://www.december.com/cmc/mag/1997/jan/december.html

Eaton, L., & Louw, J. (2000). Culture and self in South Africa: Individualism-collectivism predictions. *The Journal of Social Psychology, 140*(20), 210–217.

Farley, M. (2005, November 23). New Yorkers meet on net, play on streets. *The Seattle Times.* Retrieved March 2005, from http://seattletimes.nwsource.com/html/nationworld/2002641167_streetgames23.html?syndication=rss

Flaherty, L., Pearce, K., & Rubin, R. (1998). Internet and face-to-face communication: Not functional alternatives. *Communication Quarterly, 46*(3), 250–268.

Global Internet Statistics. (2005). *Global reach.* Retrieved September 2005, from http://www.glreach.com/globstats/index.php3

Gudykunst, W. B., Matsumoto, Y., Ting-Toomey, S., Nishida, T., Kim, K., & Heyman, S. (1996). The influence of cultural individualism-collectivism, self-construals, and individual values on communication styles across cultures. *Human Communication Research, 22,* 510–543.

Haythornthwaite, C. (2000). Online personal networks: Size, composition and media use among distance learners. *New Media & Society, 2*(2), 195–226.

Heydenfeldt, J. G. (2000). The influence of individualism/collectivism on Mexican and U.S. business negotiation. *International Journal of Intercultural Relations, 24,* 383–407.

Hofstede, G. (1980). *Culture's consequences: International differences in work-related values.* Beverly Hills, CA: Sage.

Hongladarom, S. (1998). Global culture, local cultures, and the Internet: The Thai example. *Electronic Journal of Communication, 5*(4). Retrieved October 30, 2007, from http://www.cios.org/EJCPUBLIC/008/3/008315.html

Internet Statistics Information System. (2008). *Internet population and usages.* National Internet Development Agency of Korea. Retrieved May 2008, from http://isis.nida.or.kr

Internet World Stats. (2009). Internet usage statistics: The Internet big picture. Retrieved March 2, 2010, from http://www.internetworldstats.com/stats.htm

Jeong, J. (2000, July 6). Online school reunion. *Dong A Ilbo.* Retrieved March 2004, from http://www.donga.com/fbin/searchview?n=200007060477

Katz, J. (1997, December). The digital citizen. *Wired, 5*(12), 68–78.

Kim, H., & Papacharissi, Z. (2003). Cross-cultural differences in online self-presentation: A content analysis of personal Korean and U. S. home pages. *Asian Journal of Communication, 13*(1), 100–119.

Kim, J. (2001, June 22). Criminals derived from the IRC. *Dong A Ilbo*. Retrieved March 2004, from http://www.donga.com/fbin/searchview?n=2001106220259

Kim, M. S. (1999). Cross-cultural perspectives on motivation of verbal communication: Review, critique, and a theoretical framework. *Communication Yearbook, 22,* 51–89.

Kim, M. S., Aune, K. S., Hunter, J. E., Kim, H., & Kim, J. (2001). The effect of culture and self-construals on predispositions toward verbal communication. *Human Communication Research, 27*(3), 382–408.

Kim, M. S., Hunter, J. E., Miyahara, A., Horvath, A. M., Bresnahan, M., & Yoon, H. J. (1996). Individual- vs. cultural-level dimensions of individualism and collectivism: Effects on preferred conversational styles. *Communication Monographs, 63,* 29–49.

Lee, B., Lancendorfer, K., & Lee, K. (2005). Agenda-setting and the Internet: The intermedia influence of Internet bulletin boards on newspaper coverage of the 2000 general election in South Korea. *Asian Journal of Communication, 15*(1), 57–71.

Markus, H. R., & Kitayama, S. (1991). Culture and the self: Implications for cognition, emotion, and motivation. *Psychological Review, 98*(2), 224–253.

McCroskey, J. C., & Richmond, V. P. (1987). Willingness to communicate. In J. C. McCroskey & J. A. Daly (Eds.), *Personality and interpersonal communication* (pp. 129–156). Newbury Park, CA: Sage.

McKenna, K., & Bargh, J. (1999). Causes and consequences of social interaction on the Internet. *Media Psychology, 1,* 249–269.

McLuhan, M. (1964). *Understanding media*. New York: McGraw-Hill.

Ogan, C. (1993). Listserver communication during the Gulf War: What kind of medium is the electronic bulletin board? *Journal of Broadcasting & Electronic Media, 37*(2), 177–196.

Pacey, A. (1992). *The culture of technology*. Cambridge, MA: MIT Press.

Papacharissi, Z., & Rubin, A. M. (2000). Predictors of Internet use. *Journal of Broadcasting & Electronic Media, 44*(2), 175–196.

Park, H. S., & Levine, T. R. (1999). The theory of reasoned action and self-construal: Evidence from three cultures. *Communication Monograph, 66,* 199–218.

Perse, E. M., & Ferguson, D. A. (2000). The benefits and costs of Web surfing. *Communication Quarterly, 48*(4), 343–359.

Polyorat, K., & Alden, D. (2005). Self-construal and need-for-cognition effects on brand attitudes and purchase intentions in response to comparative advertising in Thailand and the United States. *Journal of Advertising, 34*(1), 37–48.

Rapp, H. (2005, November). Flash mob brings mystery and excitement to UF. *The Independent Florida Gator.* Retrieved November 2005, from http://www.alligator.org/pt2/051121mob.php

Rheingold, H. (2000). *The virtual community: Homesteading on the electronic frontier* (Rev. ed.). Cambridge, MA: MIT Press.

Rintel, E. S., & Pittam, J. (1997). Strangers in a strange land: Interaction management on Internet relay chat. *Human Communication Research, 23*(4), 507–534.

Shin, Y. (1999). *Beongae* meeting. *Dong A Ilbo.* Retrieved February 2004, from http://www.donga.com/fbin/searchview?n=199908190242

Singelis, T. M., & Brown, W. J. (1995). Culture, self, and collectivist communication: Linking culture to individual behavior. *Human Communication Research, 21,* 354–389.

Singelis, T. M. (1994). The measurement of independent and interdependent self-construals. *Personality and Social Psychology, 20*(5), 580–591.

Somech, A. (2000). The independent and the interdependent selves: Different meanings in different cultures. *International Journal of Intercultural Relations, 24,* 161–172.

Stafford, L., Kline, S., & Dimmick, J. (1999). Home e-mail: Relational maintenance and gratification opportunities. *Journal of Broadcasting & Electronic Media, 43*(4), 659–669.

Stewart, C., Shields, S., & Sen, N. (1998). Diversity in on-line discussions: A study of cultural and gender differences in listservs. *Electronic Journal of Communication, 8*(3). Retrieved February 2004, from http://www.cios.org/www/ejc/v8n398.htm

Sun, H. (2001, October). *Building a culturally-competent corporate Web site: An exploratory study of cultural markers in multilingual Web design.* Paper presented at ACM SIGDOC 2001: Communicating in the New Millennium, Santa Fe, NM.

Thompsen, A., & Ahn, D. (1992). To be or not to be: An exploration of e-prime, copula deletion and flaming in electronic mail. *Et Cetera: A Review of General Semantics, 49,* 146–164.

Ting-Toomey, S. (1997). Intercultural conflict competence. In W. R. Cupach & D. J. Canary (Eds.), *Competence in interpersonal conflict* (pp. 121–147). New York: McGraw-Hill.

Triandis, H. C. (1995). *Individualism & collectivism.* San Francisco, CA: Westview Press.

Valkenburg, P. M., & Soeters, K. E. (2001). Children's positive and negative experiences with the Internet. *Communication Research, 28*(5), 652–675.

Varis, T. (1999). Values and the limits of the global media in the age of cyberspace. In M. H. Prosser & K. S. Sitaram (Eds.), *Civic discourse: Intercultural, international, and global media* (pp. 5–18). Stamford, CT: Ablex.

Walther, J., Anderson, J., & Park, D. (1994). Interpersonal effects in computer-mediated interaction: A meta-analysis of social and antisocial communication. *Communication Research, 21*(4), 460–487.

Walther, J. (1992). Interpersonal effects in computer-mediated interaction: A relational perspective. *Communication Research, 19,* 52–90.

Walther, J. (1996). Computer-mediated communication: Impersonal, interpersonal, and hyperpersonal interaction. *Communication Research, 23*(1), 3–43.

Weiser, E. (2000). Gender differences in Internet use patterns and Internet application preferences: A two-sample comparison. *Cyber Psychology & Behavior, 3*(2), 167–178.

Worth, S., & Adair, J. (1970). Navajo filmmakers. *American Anthropologist, 72*(1), 9–34.

Zhang, J., & Shavitt, S. (2003). Cultural values in advertisements to the Chinese x-generation. *Journal of Advertising, 32*(1), 23–33.

PART IV

Informing Pedagogy

http://dx.doi.org/10.2190/CWDC9

CHAPTER 9

Teaching Effective Technology Use in Technical and Professional Communication Programs Based in Colleges of the Humanities

Laura McGrath

OVERVIEW

Technical and professional communication (TPC) educators face the challenge of preparing students to be effective rhetoricians, critical thinkers, and, in the Digital Age, users of technologies. In its exploration of the expertise that TPC students need in order to make informed use of information and communication technologies (ICTs), this chapter advocates an approach that incorporates attention to skill building, but emphasizes technologies as tools of literacy and draws on humanities methods for investigating use practices and the rhetorical dimensions of computer-mediated communication. By providing a framework for assessing curricula, pedagogies, textbooks, and instructional spaces, this chapter encourages educators to evaluate how well they are preparing students. Arguing that the degree to which students are able to make strategic use of technology is in part a function of how that technology has been incorporated into the curriculum, the author advocates an integrated, rhetoric-based approach that supports students in becoming technical rhetoricians who engage in reflexive praxis.

We no longer have, if we ever did, the luxury of examining technologies, their practices, and their implications for the profession from a safe critical distance. Today, the work of a professional communicator is unavoidably computer mediated—and mainly, if not exclusively, digital. Many teachers of technical and professional communication (TPC) who are employed in colleges of the humanities are concerned with whether their pedagogies and curricula improve students' abilities to develop the functional, rhetorical, and critical technological literacies (see Stuart Selber's [2005] *Multiliteracies for a Digital Age* for a thorough discussion of these terms) required of professional communicators in the Digital Age. For the extent to which TPC students make meaningful use of the information and communication technologies (ICTs) pervading their workplaces after graduation is in part a function of how technology was integrated into their programs of study. By *meaningful use*, I refer to a situation in which "the user exercises a degree of control and choice over technology and content" (Selwyn, 2004, p. 352). Students' technical competencies and knowledge—part of their ability to exercise "control and choice"—have a bearing on their success in the workplace. As such, TPC programs should be taking a close look at what professional communicators need to know and do in the workplace; what and how their students are learning about technology; and how curricula can better support students in their acquisition of necessary technical competencies and thinking skills, such as critical awareness, analytical thinking, and rhetorical understanding.

Let me begin by addressing the problematic way in which digital technology often is constructed as a threatening, nonhumanistic influence. Nancy Allen and Steven Benninghoff's (2004) "TPC Program Snapshots: Developing Curricula and Addressing Challenges" makes the problem concrete. Allen and Benninghoff raise throughout that article, and especially in the section "Implications and Challenges for Tomorrow," the issue of whether TPC programs have maintained and will continue to maintain "literacy and humanities interests while incorporating more and more digital technology into their curricula" (p. 179). The authors cite Selber's concern with "how instructors would prepare their students to use technology effectively in their future professions without losing sight of literacy and humanistic goals" (1994, p. 157) and ask, "Are today's TPC programs successfully meeting the challenges of maintaining a humanities perspective while also changing to meet new demands from science and technology?" (p. 179). They conclude, with what may be a sense of relief, that survey data do show "literacy and humanities interests" continuing to drive TPC programs.

Allen and Benninghoff's application of Selber's question is problematic because it implies that attention to the effective use of technology is somehow in conflict with "literacy and humanistic goals." Perhaps, as I have already suggested, "new demands from science and technology" seem so menacing because those who fear them associate them with a vocational model that focuses on software and computer skills rather than on the critical thinking and written communication skills that typically have been at the heart of humanities-based TPC programs. Perhaps these demands are also of concern because they call into question the purpose and place of TPC programs and challenge educators to reevaluate the sufficiency of their own expertise.

It is necessary, though not necessarily easy, for educators to adapt to changing literacy tools and to their emerging critical and rhetorical uses. What is more, TPC students need to study such uses as they acquire technical skills. Assessing current curricula, pedagogies, textbooks/instructional materials, instructional spaces, faculty expertise, and student needs helps to ensure successful integration of digital technology into humanities-based professional writing curricula.

STEP 1: UNDERSTANDING WHAT TPC STUDENTS NEED TO KNOW ABOUT TECHNOLOGIES AND PRACTICES

For TPC specialists, the ability to communicate clearly, accurately, and correctly through writing remains vital. Yet the title *communicator* is often preferred over *writer* in the TPC field because it more accurately reflects the diverse responsibilities and skill sets demanded in the workplace. For example, consider Maurice Martin and Richard O'Sullivan's (2007) "The Case for 'Technical Communicator,'" which offers this updated occupational description:

> Develop and design instructional and informational tools needed to assure safe, easy, proper and complete use of technical goods. Combin[e] multi-media knowledge and strong communications skills with technical expertise to educate across the entire spectrum of users' abilities, technical experience, and visual and auditory capabilities. (pp. 7–8)

Of special note in this description are *design, multimedia knowledge*, and *technical expertise*. Martin and O'Sullivan also state that contemporary technical communicators must "produce graphic design and animation" and "analyze and select technology and media used to deliver technical information" (p. 3). In terms of technology, then, technical communicators must, in the view of these researchers, possess not only the ability to work with technologies as tools but also what I would call a rhetorical knowledge of technology. It is this rhetorical knowledge of technology that enables communicators to "analyze and select technology and media" when faced with a communication challenge.

As Allen and Benninghoff (2004) note, "TPC is a field in which the ability to use communication tools, particularly the software tools common in industry, plays a prominent role, especially for the new graduate trying to land a first job" (p. 167). Although it is difficult to predict exactly which software programs graduates will encounter in the workplace, students can certainly benefit from gaining expertise in the nearly ubiquitous Microsoft Office suite and industry-standard programs such as Dreamweaver (HTML editor), FrameMaker (XML editor), and other Adobe products.

Comment from employers amplifies what TPC students particularly need to know about technology. According to Lee-Ann Kastman Breuch (2002), her university's industry partners agreed

> that there were no specific "tools" that students should learn, for they acknowledged that tools vary widely from workplace to workplace. However, they collectively voiced the expectation that students understand technologies and have the aptitude to learn them quickly; thus, they implied that students be very familiar with common and emerging technologies in their academic programs. (p. 268)

"Common" technologies with which TPC students should be familiar might include workplace communication tools such as e-mail and instant messaging (IM), in addition to the word-processing and document-design programs already mentioned. "Common" might also refer to more advanced but relatively standard programs such as desktop publishing, HTML/XML editing software, and image manipulation, topics that Aimee Kendall Roundtree as well as Rudy McDaniel and Sherry Steward address in their chapters in this anthology. "Emerging" technologies might refer to not only new versions of software but also to the Web technologies that have been transforming use practices and the production, organization, and distribution of content (e.g., blogs, wikis, and social bookmarking tools).

Like Kastman Breuch (2002), Kenneth Rainey, Roy Turner, and David Dayton (2005) asked employers to describe the technology-related "skills and competencies" that a technical communicator needs in order to succeed in the workplace. Respondents indicated that "a person need not know all of the specific applications but should have a general familiarity with how the tools work and an ability to learn new tools quickly" and that "the ability to adapt to new situations and to learn new software quickly is far more important than knowledge of specific software packages" (p. 333). The two most important technical competencies cited by the employers studied were the "ability to assess and to learn to use technology" and the "ability to operate word-processing and document-design applications" (p. 328). Put another way, *technical aptitude* is one important component of overall *technical competency*, which also includes specialized expertise with, for example, word processing and document design. In terms relevant to TPC students, *technical aptitude* refers to a general familiarity with technology and the ability to learn new tools and adapt to changing digital communication environments.

The level of specialized technical expertise TCP students should acquire is a matter of debate, so it may be counterproductive for students to spend too much time mastering specific software programs. In contrast, students undeniably can benefit from "instruction in actively learning how to learn new technologies" (Rainey et al., 2005, p. 325), for most employers are willing to train promising candidates who demonstrate general technical aptitude.

If the development of technical aptitude is to occur as acquisition of rhetorical knowledge of technology, it seems especially essential for TPC educators to encourage students to use technologies in light of audience and situation. Writing about Isocrates's educational program, Takis Poulakos (1997) observed, "Because knowledge itself cannot suggest the uses to which it can be put, it follows that instruction in a field of study will always be incomplete unless it cultivates the aptitude to know what to do with and how to use acquired knowledge" (p. 96). In the context of this chapter, it could be said, then, that TPC educators should focus not only on helping students acquire technical aptitude but also on helping them acquire within a rhetorical framework "the aptitude to know what to do with and how to use" what they learn.

This point is supported when one considers the relevance of Web 2.0 or social media to the field of TPC. A 2006 *Communication World* article suggests that at a time when the promise of the read/write Web has come to fruition and users are interacting with and producing content as never before, "organizations that wish to wield influence must do so through engagement with the audience. They must participate in the conversation. The days of crafting and delivering a top-down message are dwindling" (Holtz, p. 25).

Employers offer comment on this shift as well, as the following anecdote suggests. After a recent conversation with a practitioner at an International Association of Business Communicators meeting, a colleague approached me about developing a social-media course for the applied-writing students in our university's Master of Arts in Professional Writing program. The practitioner had mentioned that her firm would not hire professional communicators who did not have background in social media (e.g., working knowledge of blogs, social networking applications like MySpace and Facebook, user-generated content sharing sites like YouTube and Flickr, tagging/folksonomy, etc.). In a follow-up e-mail, the practitioner went on to say that "most companies are looking for communications professionals who understand how to build online communities using multiple social media tools that facilitate consumer-generated content development" (Marilynn Mobley, personal communication, September 27, 2007).

Students of TPC may be immersed in social media as users and in this way develop technical aptitude. If they are not, they can be encouraged to explore these tools as platforms for communication within various professional communication contexts. Above all, students need guidance in thinking rhetorically about use practices, affordances and limitations, social and ethical implications, and the impact of social media on the field. This is because social media disrupt the traditional communication model by which a writer creates a static information product for a user. William Hart-Davidson (2007) emphasizes that "Web 2.0 represents not just a shift in user behavior, but also a shift in the power of users to shape their own experiences and, thereby, to shape the experience of others" (p. 12). Students of TPC need to be prepared to respond strategically to these changes in user behaviors, expectations, and options. Technical aptitude and rhetorical knowledge combine in a powerful way when students are challenged to apply what they know and at the same time to consider the implications of their choices with regard to, for example, issues of authority and values.

In addition to developing (a) technical aptitude with and rhetorical knowledge of social media and (b) an understanding of their own role in technology-mediated communication environments, TPC students need to gain expertise in working with *more than* words. Rich Rice and Carole Clark Papper (2005) propose that attention to multimodality "should underpin the delivery and content of each assignment" (pp. 296, 303). Because multimedia content most frequently is created with digital tools and distributed either online or on a CD or DVD, students may benefit from an introduction to the use of, for example, cameras, recording devices, and image and audio software.

And yet, again, effectively gathering, shaping, and delivering information requires skillful application of rhetorical knowledge. As with social media, multimodality can be examined effectively in TPC courses incorporating the intellectual apparatuses of the humanities scholar. As students develop technical aptitude within a critical and rhetorical framework, they will gain an understanding of when, why, and how to use various specific tools to accomplish communicative goals.

Every time TPC students are faced with communication tasks, their deliberations about audience, purpose, and context should account for technologies, interfaces (points of contact), and modes. Ideally, rhetorical knowledge of technology will lead to such praxis. (Here, I am drawing on and repurposing Selber's suggestion that the objective of rhetorical literacy within a computer literacy program is reflexive praxis [2005, p. 25].)

In sum, current TPC students need to have at least a basic understanding of the following:

- computer technologies,
- strategies for learning how to use new hardware and software,
- common communications technologies,
- emerging communications technologies and use practices,
- multimedia design, and
- rhetorical knowledge of critical approaches to technology.

This list can serve as a starting point for conversations aimed at its refinement. Further, it can be used by TPC educators and program administrators as they assess how effectively their students are being prepared to make informed and meaningful use of pervasive information and communication technologies in professional work.

STEP 2: ASSESSING PROGRAMS

Once TPC educators have a clear sense of what future professionals need to know about technologies and their use, they can begin to assess curricula, pedagogy, textbooks/instructional materials, instructional spaces, and faculty preparedness and expertise. Programs that have been keeping pace with change or even leading change can benefit from the ways in which assessment encourages teachers to articulate our core values and to justify our choices. The assessment process may be most useful to teachers or program directors who have received a call to action (as I did in the social media scenario) and who, already having recognized a need for improvement, have turned to this chapter and other resources for guidance. The assessment framework outlined in this section can be used by programs or by individual TPC teachers.

In its potential to surface areas of strength and of lag in TPC curricula, assessment also enables programs to identify and address divergences in students' knowledge production as engendered by digital technologies. Neil Selwyn (2004) figures these differences along a continuum that should be seen "in terms of the information, resources, applications and services that individuals are accessing via new technologies." Selwyn emphasizes "[effective] access in practice over theoretical (or formal) access" to information and communications technology (p. 347) and asserts that the fundamental issue is one of effective use once access has been achieved.

According to Selwyn, effective use requires (a) *cultural technological capital*, or formal and informal training and "socialization into technology use and 'techno-culture'" and (b) *social technological capital*, or "networks of 'technological contacts' and support" (p. 355). Selwyn's points about cultural and social technological capital and effective access are relevant to TPC educators involved in program assessment. This is so because, in Selwyn's view, effective access is tied to meaningful use of ICTs, which depends on a variety of activities, including production, political activity, social activity, and consumption. To make meaningful use of ICTs, an individual needs not only access to but also training, socialization, and support in these technologies.

Training, socialization, and support should be integrated in TPC programs so as to foreground the connections between what might be called "ICT literacy" and "rhetorical and critical literacy." The main focus of an integrated approach would not be on helping students develop technology-related job skills but instead would be on helping them become literate professional communicators. This way of thinking about ICT literacy as one essential part of professionalization can appropriately guide TPC educators' assessment efforts. In the section that follows, the concept of *meaningful use* has been incorporated into many of the guiding questions.

STEP 3a: ASSESSING CURRICULA AND PEDAGOGY

As I have argued, current TPC students need to develop

- technical competency, including both general technical aptitude and a degree of specialized expertise;
- rhetorical knowledge and critical awareness of technology, and specifically of the ability to apply the apparatuses of the humanities scholar and of the rhetorician to the study of common and emerging technologies, their impacts on the field, and their social and ethical implications generally; and
- the ability to "gather, shape, and deliver" content in a variety of modes (Rice & Papper, 2005, p. 303) and to use available technologies responsibly and effectively to achieve communicative goals.

Curricula and pedagogies, then, should prepare future specialists to be, among other things, what is sometimes referred to in the context of discussions of new media as *multiliterate*. Recognizing this need, Kelli Cargile Cook (2002) offers a framework for technical communication pedagogy that may be a useful tool for assessing TPC curricula. She identifies the following as characteristics of technological literacy:

- a working knowledge of technologies that helps professional communicators to produce communications, documents, or products;
- an awareness of how these technologies promote social interactions and collaboration;
- an ability to research how users work with technologies; and
- an ability to critique this research and act upon it to make decisions and produce documents designed with and for users. (p. 13)

Assignments and courses designed to facilitate critical thinking about technology as well as reflexive praxis are key to helping students achieve multiliteracies. Richard Selfe's (2007) chapter on technological activism in *Resources in Technical Communication* proposes an assignment sequence that focuses on "technology-rich communication environments" and engages students in research, collaboration, synthesis, analysis, and presentation. The purpose of the assignment is to expand students' "understanding about what goes on in [technology-rich communication environments]" and to get the students involved in "making suggestions for change that would serve a community of communicators" (p. 148). Our challenge, as teachers, is to support the development of multiliteracies throughout the curriculum and to integrate technology-related thinking and doing skills in such a way that these considerations become a standard part of all communication tasks.

The graduate technical communication course in information architecture that Michael Salvo (2002) has described also models how to engage students as technical rhetoricians, in working "with advanced digital technologies, [articulating] their relationship with design, [working] collaboratively with classmates . . . [and articulating] the unfamiliar relationships they had with technology and the responsibilities they would have both to users and technology producers" (p. 323). Thinking about technical and professional communicators as technical rhetoricians and encouraging TPC students to define themselves this way would seem very important. But, once again, such approaches and concepts are useful only insofar as they are integrated into a TPC student's program of study.

Indeed, the problem with proposing an ever-increasing number of specialized courses in order to "address" technology hinges on the issue of integration. It may seem easier to meet emerging needs and address new theories and practices in specialized courses, and there is something to be said for the kind of sustained immersion such courses can provide. In addition to offering such courses, however, TPC educators should find ways to integrate attention to technologies and use practices across the curriculum and to offer regular opportunities for reflexive praxis.

Guiding questions for assessing curricula and pedagogies might include the following:

- Is attention to the development of rhetorical knowledge and critical awareness of technology integrated across the TPC curriculum?
- How do current pedagogical approaches to technology promote or fail to promote rhetorical awareness and meaningful use? Is reflexive praxis emphasized?
- Are common and emerging technologies and use practices being addressed in ways that encourage students to consider their impact on the field and their ethical and social implications, including how they might "promote social interactions and collaboration" (Cook, 2002, p. 13)?
- Are students offered opportunities to develop technical competency in ways that emphasize general technical aptitude but also promote the development of specialized expertise in, for example, word processing and document design?

STEP 3b: ASSESSING TEXTBOOKS AND INSTRUCTIONAL MATERIALS

Textbooks and other instructional materials (e.g., online supplements, CDs, DVDs) used in TPC courses are shaped by their authors' expertise, values, and ideologies, with an obvious tendency to shape student learning. A review of available TPC books and materials is beyond the scope of this chapter and would become outdated rapidly. I will instead focus attention on key issues for TPC educators to consider when assessing and selecting textbooks.

Tools and use practices evolve rapidly, so one challenge is finding material that is current enough to provide students with an accurate understanding of how technical and professional communicators use technical tools, create and manage digital information for various audiences, and make technology-related decisions in the workplace. And it is also necessary to look for discrepancies between textbook content and current workplace practices. The need for up-to-date instructional material means that TPC educators will need to schedule textbook assessment on a regular basis, ideally once per academic year. Consulting publishers' lists, dividing up the most promising options among TPC faculty, and then sharing informal reviews is an efficient approach.

A second challenge is to find books and materials that provide sufficient, well-integrated attention to—and here I will once again use Selber's (2005) terminology—functional, critical, and rhetorical technological literacy. Cook's (2002) assertion that "many introductory technical communication textbooks continue to isolate instruction in certain literacies into individual chapters rather than integrating these literacies throughout the course of study" (pp. 6–7) continues to be a relevant critique. A textbook may "isolate instruction in" technological literacy, for example, by relegating all attention to technology to sidebar items or to "how to" sections. In such scenarios, technology tends to be addressed in terms of functional literacy alone, and content that encourages students to think critically about technology is lacking, as is attention to the rhetorical aspects of technology use in TPC. Classroom instruction can certainly make up for some deficiencies, but textbooks that isolate technological literacy instruction undermine rather than complement attempts to help students attend to the rhetorical and critical aspects of technology use.

In contrast to such textbooks are those that approach technology use from a rhetorical perspective; and it is this sort of textbook that may best support TPC educators' efforts in the classroom. One example worth mentioning is Sidney Dobrin, Christopher Keller, and Christian Weisser's (2008) *Technical Communication in the Twenty-First Century*, which was "written with [ethics and technology] in mind from the outset—as should be the case for all . . . workplace writing—rather than addressing them as secondary considerations" (p. iv). Further, the authors suggest that "students should understand technology as an integral part of rhetorical thinking and writing that shapes how and what they write" (p. 6).

As indicated in the text's prefatory material, relevant technologies and related issues are covered throughout the chapters. Dobrin et al.'s text provides students with a five-step problem-solving approach (plan, research, draft, revise, distribute) that draws attention to technology as a practical, rhetorical consideration that is an important part of each communication task. The text also provides up-to-date information about tools and software with which technical communicators should be familiar. Chapter 3, "Technical Writing and Electronic Technologies," which is devoted to exploring technology and workplace writing, covers word processors, Web-authoring software, graphics and imaging software, desktop publishing software, help-authoring tools, single-sourcing programs, the Internet, e-mail, the World Wide Web, electronic messaging, videoconferencing, and groupware. With such topics, a textbook like *Technical Communication in the Twenty-First Century* can support TPC educators' efforts to promote an integrated, rhetoric-based approach to technology.

The best TPC textbooks will encourage students to see themselves as technical rhetoricians as well as writers of standard technical and professional documents. Instead of treating technology as an aside or in a limited and limiting way, excellent textbooks and instructional materials encourage students to examine technologies and use practices as part of a problem-solving process.

Guiding questions for assessing textbooks' and instructional materials' approaches to technology might include the following:

- Is attention to technology as an element of the communicator's rhetorical situation integrated throughout the material?
- Does the material include rhetoric-based discussions of technology in addition to tool or skill "tips?" Does it encourage students to think about the implications for practitioners working in the field as well as the potential social and ethical implications?
- Is the technological information up to date and reflective of common and emerging use practices? Does it support the development of technical competency (aptitude and expertise)?

STEP 3c: ASSESSING INSTRUCTIONAL SPACES

Assessment should also include careful consideration of the instructional spaces in which TPC students work and learn. Whether these spaces are physical or virtual, the resources that they contain, the way resources are arranged, and the structures that support, maintain, upgrade, and control access to resources should be assessed thoughtfully. For discussions of infrastructure and computer classrooms that might provide a useful background for assessment of on-ground instructional spaces, see Dànielle Nicole DeVoss, Ellen Cushman, and Jeff Grabill (2005), Cynthia Selfe (1989), Richard Selfe (2004), Kathleen Welch (2005), the WIDE Research Center Collective (2005), and Sean Williams (2002).

Although only Williams' and Welch's articles, both published in *Technical Communication Quarterly*, are written for TPC educators, all these resources offer valuable guidance. The WIDE piece, for example, offers two examples of "technology-rich instructional space[s] that . . . facilitate the tasks typical of a writing classroom" (Changed spaces . . . section, para. 9), examples that may be useful for comparison purposes or as models of what other institutions might consider doing with their own instructional spaces.

Guiding questions for assessing instructional spaces might include the following:

- Are instructional spaces conducive to explorations of the rhetorical possibilities of various tools and immersion in digital communication environments?
- What tools are available in the instructional spaces? Are any essential tools (hardware or software) absent, unavailable, in disrepair, or out of date?
- Are the structures that support and control access to these spaces limiting or enabling?

STEP 3d: ASSESSING FACULTY (SELF-ASSESSMENT)

As I have mentioned, facing the "new demands from science and technology" challenges TPC educators to reevaluate the sufficiency of their own expertise. Faculty need to know what TPC students need to know about technology. That a number of TPC faculty are not practitioners, have never been practitioners, or have not been practitioners for many years presents a challenge; but this simply means that TPC educators must make a professional commitment to staying informed about technologies, their practices, and industry expectations. As humanists and rhetoricians, scholar-educators should be adept at discerning emerging practices of significance and making sense of these practices on a theoretical level.

The argument that programs may "need to add faculty who specialize in the areas the program intends to target" (Allen & Benninghoff, 2004, p. 176) can be problematic if addition of specialists is undertaken in lieu of integrating a rhetoric-based approach to technology across the TPC curriculum. Attention needs to be paid first to current faculty, their professional development needs, and the potential benefits of strategic partnerships among them, their colleagues in other disciplines, and practitioners. Along these lines, assessment of faculty and self-assessment might inventory expertise, attitudes, needs, and connections.

Guiding questions for assessing faculty or for self-assessment might include the following:

- What technological expertise do faculty currently possess? Is this expertise limited to the use of tools, or does it include expertise in rhetorical and critical approaches to technologies and use practices?
- How do faculty use technology in their work as teachers and scholars? If faculty are also or have recently been practitioners, how do/did they use technology in the workplace?
- What attitudes toward or beliefs about technology do faculty possess?
- What do faculty have to say about what they need in terms of access, support, professional development, etc.?
- Do faculty have ties to allied fields or to industry that might be useful to pursue?

STEP 4: CONSIDERING STRATEGIES FOR BETTER PREPARING TPC STUDENTS

Once assessment of curricula, pedagogy, instructional materials, instructional spaces, and faculty is complete, TPC educators will have a better sense of where they are and where they want to go. From this vantage point, individuals or programs can begin to consider strategies for better preparing TPC students to be effective, strategic users of information and communication technologies.

Situational factors—resource allocation, programmatic goals, campus politics, and so on—differ dramatically across institutions, so it is impossible to generalize "solutions" for improving TPC education although it may be useful to think in terms of "promising strategies." First, I propose that TPC educators adopt the integrated, rhetoric-based approach that has been advocated throughout this chapter. Second, depending on what assessment reveals about needs, strengths, and weaknesses, educators may also want to consider the potential advantages of interdisciplinary initiatives and strategic partnerships with industry.

Saul Carliner (2005) begins "Integrating the Web into Education for Technical Communication Majors" with a statement that speaks to the changing nature of the field:

> What has changed is that many people in the field [now] realize that what we do is much larger than writing and editing and that the fields that can help us think about what we're doing are really diverse: Cognitive psychology, graphic design, typography, human factors, comprehension, management and planning. I think that more people today have to be a jack of all trades even if they're working in large organizations where there's lots of support. (p. 263)

Combine those diverse knowledge requirements with what TPC students need to know about writing, rhetoric, and common and emerging technologies, and Carliner's is a tall order for any program to fill. This is especially true when we consider the differing levels of faculty expertise with technology.

How, as Christine Abbott (2004) asks, "do we improve the quality of our programs and give students meaningful education experiences, without substantial additional resources and without putting further pressure on ourselves?" (p. 255). Interdisciplinary initiatives and strategic partnerships with industry may help by engaging multiple stakeholders—faculty, students, industry affiliates—in mutually beneficial collaborations.

Interdisciplinary Initiatives

In "What Do We Need to Teach About Knowledge Work?" Clay Spinuzzi (2006) describes the growing importance of horizontal learning, which, in contrast to the "rigid hierarchical separations" of the vertical learning long standard in higher education, "involves crossing field, trade, and disciplinary boundaries in order to collaborate and develop crossfunctional, interlinked work" (para. 7). As Spinuzzi explains, in the Digital Age "workers from historically separated activities suddenly must interact, collaborate, and learn enough of each others' social languages and genres to work together" (para. 3). Thus, by becoming involved in interdisciplinary collaborations during their academic careers, TPC students may be better prepared knowledge workers when they enter the workplace. The Transcriptions project at the University of California at Santa Barbara exemplifies the potential of such a collaboration:

> Transcriptions uses the themes and practices of information technology to create research activities that bridge not just between academic disciplines but between the academy and other professional sectors of society. Faculty and students in the project, for example, collaborate or visit with members of such other departments on campus as Art Studio, Media Arts & Technology, Film Studies, or Psychology. . . . Project participants also meet with or interview people from the worlds of business and private-industry information technology. (Liu, 2002, para. 17)

It is not hard to imagine how TPC educators, colleagues in other departments (e.g., art, computer science, psychology), students, and industry affiliates might gain knowledge and opportunities for reflexive praxis and horizontal learning through creative interdisciplinary work with a technological focus.

Strategic Partnerships with Industry

Like interdisciplinary initiatives, strategic partnerships with industry may help stakeholders—faculty, students, and industry affiliates—learn together. Because they allow exposure to situated practice, strategic partnerships with industry can be especially useful in fleshing out TPC students' technical knowledge and skills and can succeed in concert with the integrated, critical, and rhetorically focused approach to technology I have advocated in this chapter.

Many approaches to strategic partnership are possible. Some focus on connecting educators with practitioners. Abbott (2004) suggests taking advantage of "opportunities for collaboration" such as Society for Technical Communication-sponsored "faculty internships in corporations" (p. 255) and imagines "a virtual consortium of top-notch professors, teachers, scholars, and practitioners collectively planning, teaching, conferencing, and mentoring in a cybercollaboratorium" (p. 268). Partnerships such as the ones envisioned by Abbott can open the lines of communication between the academy and industry and help faculty stay in touch with the latest communication challenges, technologies, and practices.

Other approaches focus on connecting students and practitioners. In "The Impact of the Internet and Digital Technologies on Teaching and Research in Technical Communication," Laura Gurak and Ann Duin (2004) suggest that "programs can partner with industry to offer co-created courses that provide students with 'just in time' experiences with up-to-date resources" (p. 189). Even an internship can be considered a strategic partnership when students are asked to filter knowledge gained during the program back into the academy, perhaps through symposia or colloquia that also bring together faculty and practitioners/employers. Faculty can then guide student presenters and their audiences into discussions that foreground the rhetorical, social, and ethical dimensions of situated practices.

Changes in the field of TPC seem to call for us to reach outward in these ways and to establish connections with individuals in other disciplines and fields. Educators whose programs are still quite vertical and discipline-bound may determine that both interdisciplinary initiatives and strategic partnerships are feasible and appropriate; or their situation may call for only one or neither of these approaches. Regardless, the move outward can help educators face the challenge of staying ahead of the curve. If connected to opportunities to develop rhetorical knowledge and critical awareness of technology, educators may find that the approaches described provide students with information, skills, and experience that will serve them well, beyond the classroom.

CONCLUSION

Technical and professional communication teachers working in colleges of the humanities are confronted with many potential paths into the Digital Age. I have chosen to focus on a multilayered approach that acknowledges the need to prepare students to make meaningful use of the information and communication technologies of contemporary workplaces. This approach incorporates attention to skill building but emphasizes technologies as tools of literacy and draws on humanities methods for investigating use practices and the rhetorical dimensions of computer-mediated communication. I have focused specifically on ways that educators might assess and improve their programs and support TPC students in becoming technical rhetoricians who engage in reflexive praxis.

Although the New Media Consortium's (2007) *Horizon Report* states that "faculty need curricula that adapt to the pace of change and that teach the skills that will be needed—even though it is not clear what all those skills may be" (p. 5), such remarks can feel paralyzing. How can we proceed when, to borrow Allen and Benninghoff's (2004) phrasing, we are standing on shifting sands? As I have attempted to show, an integrated, rhetoric-based approach to technology, combined with outward-reaching collaborative and strategic partnerships, may be helpful to both faculty and students who need to stay engaged with developments in information and communication technologies. Ongoing assessment of curricula, pedagogies, textbooks, instructional spaces, and ourselves as teachers is also essential, for such work helps us to remain mindful of our core values, weaknesses, and strengths.

REFERENCES

Allen, N., & Benninghoff, S. T. (2004). TPC program snapshots: Developing curricula and addressing challenges. *Technical Communication Quarterly, 13*(2), 157–185.

Abbott, C. (2004). (Re)Connecting theory and practice. In T. Bridgeford, K. S. Kitalong, & D. Selfe (Eds.), *Innovative approaches to teaching technical communication* (pp. 254–271). Logan: Utah State University Press.

Carliner, S. (2005). Integrating the Web into education for technical communication majors: A process-oriented approach. In C. Lipson & M. Day (Eds.), *Technical communication and the World Wide Web* (pp. 263–283). Mahwah, NJ: Lawrence Erlbaum.

Cook, K. C. (2002). Layered literacies: A theoretical frame for technical communication pedagogy. *Technical Communication Quarterly, 11*(1), 5–29.

Dobrin, S., Keller, C., & Weisser, C. (2008). *Technical communication in the twenty-first century*. Upper Saddle River, NJ: Prentice Hall.

DeVoss, D. N., Cushman, E., & Grabill, J. T. (2005). Infrastructure and composing: The when of new-media writing. *College Composition and Communication, 57*(1), 14–44.

Gurak, L. J., & Duin, A. H. (2004). The impact of the Internet and digital technologies on teaching and research in technical communication. *Technical Communication Quarterly, 13*(2), 187–198.

Hart-Davidson, W. (2007, September–October). Web 2.0: What technical communicators should know. *Intercom,* 8–12.

Holtz, S. (2006). Communicating in the world of Web 2.0. *Communication World, 23*(3), 24–27.

Kastman Breuch, L. K. (2002). Thinking critically about technological literacy: Developing a framework to guide computer pedagogy in technical communication. *Technical Communication Quarterly, 11*(3), 267–288.

Liu, A. (2002). About Transcriptions. *Transcriptions: A digital humanities project on the cultures of information*. Retrieved November 15, 2007, from http://transcriptions.english.ucsb.edu/about/

Martin, M., & O'Sullivan, R. (2007). The case for "technical communicator." *Society for Technical Communication*. Retrieved November 15, 2007, from http://www.stc.org/PDF_Files/caseTC.pdf

New Media Consortium. (2007). *Horizon report 2007*. Retrieved November 15, 2007, from http://www.nmc.org/pdf/2007_Horizon_Report.pdf

Poulakos, T. (1997). *Speaking for the polis: Isocrates' rhetorical education*. Columbia, SC: University of South Carolina Press.

Rainey, K. T., Turner, R. K., & Dayton, D. (2005). Do curricula correspond to managerial expectations? Core competencies for technical communicators. *Technical Communication, 52*(3), 323–352.

Rice, R., & Papper, C. C. (2005). Moving beyond "text only" pedagogy: Oral, print, and electronic media in technical communication assignments. In C. Lipson & M. Day (Eds.), *Technical communication and the World Wide Web* (pp. 295–304). Mahwah, NJ: Lawrence Erlbaum.

Salvo, M. (2002). Critical engagement with technology in the computer classroom. *Technical Communication Quarterly, 11*(3), 317–337.

Selber, S. (1994). Beyond skill building: Challenges facing technical communication teachers in the computer age. *Technical Communication Quarterly, 3*(4), 365–390.

Selber, S. (2005). *Multiliteracies for a digital age*. Carbondale, IL: Southern Illinois University Press.

Selfe, C. (1989). *Creating a computer supported writing facility: A blueprint for action*. Urbana, IL: NCTE.

Selfe, R. (2004). *Sustainable computer environments: Cultures of support in English studies and language arts*. Cresskill, NJ: Hampton Press.

Selfe, R. (2007). Technological activism: Understanding and shaping environments for technology-rich communication. In C. Selfe (Ed.), *Resources in technical communication: Outcomes and approaches* (pp. 145–159). Amityville, NY: Baywood.

Selwyn, N. (2004). Reconsidering political and popular understandings of the digital divide. *New Media & Society, 6*(3), 341–362.

Spinuzzi, C. (2006, September 25). *What do we need to teach about knowledge work?* Computer Writing and Research Lab White Paper series. Retrieved November 15, 2007, from http://www.drc.utexas.edu/research/what-do-we-need-teach-about-knowledge-work

Welch, K. E. (2005). Technical communication and physical location: Topoi and architecture in computer classrooms. *Technical Communication Quarterly, 14*(3), 335–344.

WIDE Research Center Collective. (2005). Why teach digital writing? *Kairos, 10*(1). Retrieved February 10, 2006, from http://english.ttu.edu/kairos/10.1/binder2.html?coverweb/wide/index.html

Williams, S. D. (2002). Why are partnerships necessary for computer classroom administration? *Technical Communication Quarterly, 11*(3), 339–358.

http://dx.doi.org/10.2190/CWDC10

CHAPTER 10

Technical Communication Pedagogy and the Broadband Divide: Academic and Industrial Perspectives

Rudy McDaniel and Sherry Steward

OVERVIEW

Technical communication is a field poised to benefit substantially from the high-speed networks, distributed workplaces, globalizing economies, and innovative communication technologies that characterize the Broadband Age. Communication today is more rapid, more complex, and more enhanced with multimedia capabilities than ever before. Interdisciplinary professionals who can navigate these high-tech networks and disseminate information to a wide range of audiences are in great demand. Regrettably, technical communicators are often ill equipped to excel in high-tech environments; as a result, the Broadband Age is contributing to the rift between technical communication pedagogy and industrial practice. In this chapter, we identify and explore various gaps that have emerged between communication practices in academe and industry. We conclude by foregrounding pedagogical activities designed to prepare technical communication students to act as leaders in the information economy.

According to *A Nation Online: Entering the Broadband Age*, the largest sector of growth in home Internet use in the United States has been aided by the transition from dial-up to broadband technology (cited in U.S. Department of Commerce, 2005). Such movement, however technically impressive, has two problematic implications in terms of digital divergence: For one, it widens the gap between those without any form of Internet access and those who now have high-speed access; for another, it creates a new gap between those with low-speed and those with high-speed access. Not surprisingly, the shift to higher bandwidth also has important consequences for the profession of technical communication. We argue in this chapter that differences in broadband usage constitute a new challenge to the at-times fraught relationship between academics teaching technical communication and practitioners working in industry.

We explore in the first part of this chapter the problems associated with pedagogical models and industrial workflows designed more for static media and documentation than for the newly interactive environment made possible by broadband technologies. This resource gap is noteworthy in the business world and has led some writers to suggest the existence of a broadband divide between large and small businesses (Arbori & Ordanini, 2006, p. 84). We will show that a similar fissure has opened between many technical communication education programs and the businesses hiring from such programs. Currently, many curricula may not be preparing students adequately for the technological and political challenges they will encounter as designers and facilitators of complex information in the highly visual and interactive environments made possible through broadband technologies. Whereas academic training in the development of critical thinking skills (skills such as audience analysis, evaluation, and assessment of source materials; selection and arrangement of materials; and other rhetorical strategies for document design) is undeniably necessary, problems may arise when graduates find themselves unprepared to work in an industry where competencies are normally assessed from within a positivist framework (cf. MacDonald, 2002; Scollon, 2003). Carolyn Miller (1979) describes this framework as a knowledge model heavily reliant on sensory data and empirical verification (p. 612). Relying on quantitative data, logical decision making, and processing rules, information technology (IT) drives the operational practices of industrial technical communication. Everything from the structure of file servers to the availability of word processing software to the policies and procedures associated with electronic mail influences the social and professional practices of industrial workflow.

Because the positivist knowledge paradigm persists throughout industry, we argue that technical communicators in the workforce tend to be classified according to technical skills and knowledge instead of writing expertise or rhetorical acumen. As a result, technically competent communicators tend to be located at the top of the institution's professional hierarchy. Technological skills and conceptual understanding of interactive media design and integration are made even more important by new broadband technologies, for these technologies enable the convergence of sophisticated media types such as video and animation. Graduates who are not technologically savvy are thus often at a disadvantage, despite their training in critical thinking and rhetoric. The notion that the person doing the thinking is more important than the tool or system being used (see Rutter, 1991) is all too often honored only when the person focuses his or her thoughts on tools and systems.

While much modern communication theory suggests that meaning is shaped by cultural and sociopolitical ideologies, ideologies that should often be called into question, examined, or even challenged, industrial logic tends to operate under a different paradigm. A central industrial belief is that processes and workflow deliverables should be *fast, efficient,* and *maintainable*. Calling into question political practices is often a risky move, especially for entry-level practitioners such as those who have just graduated from technical writing programs. Because of this dissonance between academic and industrial perspectives, recent technical communication graduates working in industry may be startled to see the emphasis on standardization, encapsulation, and IT in the creation of texts for specialized training or the distribution of technical content. This problem is exacerbated as digital systems evolve in terms of growth capability and bandwidth allowances, since many of the cutting-edge technologies designed to improve speed, efficiency, and maintainability are not available in cash-strapped universities, where professors are unlikely to be trained in the use of the latest technological tools. Students graduating from such programs may be surprised to see how different the practitioner's world is from the academic's. In industry, the pace is faster, the tools are more complex and sophisticated, and information moves very quickly through both sociopolitical (e.g., departments) and technological (e.g., optical fiber) networks. Broadband network access in this context accentuates epistemological differences and can place recent graduates in a sink-or-swim situation in which they must rapidly adapt their skill sets.

To support our argument that differences in broadband usage are influencing the field of technical communication, we consider examples from a typical career field: Simulation and training. Using several examples taken from this industry, we discuss how professional communicators are being trained in the classroom and how they are expected to function in the workplace. Next, to address some of the problems associated with these methodological differences, we propose a hybrid curricular model for assessing and producing emergent forms of interactive technical documentation, a model relying on both academic and industrial practices for support. A curricular model acknowledging both core technological competencies and professional paradigm differences can help technical communication students trained to approach knowledge through a discursive and constructivist lens to sustain themselves in environments where positivist notions hold sway. We will begin by considering the issue of technological training for technical communicators.

ACADEMIC PERSPECTIVES

Technological differences between academic and industrial practices are nothing new. In fact, these differences are often thought to be advantageous to students and soon-to-be junior employees. Traditionally, college-educated workers have learned to think and to solve problems rather than to use specific tools. This emphasis is often held up as a point of pride in the university. Noting that specific technologies are always changing, but that an ability to solve problems creatively and to feel comfortable with learning new technologies will serve an individual well across a wide range of career paths, professors might characterize an in-depth knowledge of a specific computer technology as a suitable goal for a trade/vocational school but not for a university. As a result, upon graduating from college, students may expect to move from one problem domain to another with minimal discomfort or anxiety and may assume that learning new tools should likewise be easy enough. After all, new graduates should have the ability to transfer basic theoretical skills to the design and construction of applied technologies and should be able to learn to apply these technologies to the solution of specific problems.

Certainly, the rhetorical and critical skills learned in technical communication classrooms are adaptable and configurable to a multitude of industrial tasks, as Aimee Kendall Roundtree demonstrates in this anthology in her chapter on the urgent need for more rhetorically informed application of extensible markup language (XML). As technologies continue to mature and applications become more media intensive, however, the gap between digital competencies required in industry and the basic computer skills learned in academic programs widens precipitously. In this context, rhetorical skills, critical maturity, and general knowledge of how to learn technology may be insufficient to overcome even the entry-level barriers for employment in certain industries.

This situation indicates a deepening rift between academics and industry practitioners, a rift caused in part by the fundamentally different ways in which their respective communities see the world. Academics often prefer to study theoretical issues; this preference is not necessarily surprising, as promotion and tenure committees often value theoretical research over applied productions. Even applied research in academe can tend toward the abstract, as evidenced by the use of systematic research to generalize results from a test case to a larger population. Industry practitioners, on the other hand, must continually be concerned with technical minutiae, production deadlines, and deliverables, as well as with mastering and applying the latest technologies in order to stay ahead of the fierce competition brought on by globalization. Let us now consider these differences in light of the changes introduced by broadband technology.

MULTIMODAL PRODUCTION: A DIVISIVE APPROACH?

As Gunther Kress and Theo Van Leeuwen (2001) write, "We live in a world where discourse, design, and production no longer form a unity, where teachers are trained to teach without any reference to what they are teaching, managers to manage without any reference to what they might be managing, interviewees to being interviewed without any reference to what the interviews might be about" (p. 9). Rather than diminishing this sense of postmodern disconnection, Internet and broadband technologies may be deepening it. Complex information presses from multiple sides and angles, threatening information overload through cognitive burden or distraction. To make matters worse, many of the technologies that are important for contemporary technical communication in industry (single sourcing, XML, database-driven Web design, extreme programming, Rich Internet Applications, etc.) continue to grow in complexity and sophistication as greater affordances are provided by faster technologies with wider bandwidths. Even traditional ideas about writing (e.g., the craftsman model) are threatened in this fast-paced drive catalyzed by IT-centric organizations. Thus the gap between academic and industrial thought and practice widens.

In considering the historical factors leading to this disconnect, it is useful to note how both groups traditionally have positioned themselves in regard to production processes. Kress and Van Leeuwen's discourse model, which traces communication through the practices of discourse, design, production, and distribution, enables techniques for both the articulation/analysis and the creation/invention of materials functioning in multimodal environments. Historically speaking, academic programs in technical and professional communication have tended to focus almost exclusively on the discourse and design domains, or *strata*—to use Kress and Van Leeuwen's term (p. 4), and mostly on the discourse domain, which is most abstract. In contrast, industrial workflows have been much more concerned with the highly specific production-oriented and distributive acts of communication.

The academic phase of a technical communicator's training therefore deals more with conceptual reasoning and abstract design than with production and distribution. To be sure, there are exceptions, such as the many technical communicators trained in disciplines such as computer science, engineering, or any academic field other than English studies or communication; and within these latter fields the scattering of courses concerned with the production of specialized forms of documentation such as computer software manuals or training materials. More than 85% of technical communication programs are structured around English, communication, or other departments in the humanities, however (Harner & Rich, 2005); thus, the tendency of such programs to devote large amounts of time to abstract concepts and to analyses of existing examples of communicative modes and media is not surprising. For the most part, technical communication programs focus on the learning outcomes most often associated with good *writing*, and good writing in the *general* sense—basic communication skills, and on the theoretical paradigms supporting or cultivating such skills. Programs' learning outcomes often include skills such as identifying and addressing an audience, describing procedural tasks with appropriate scope and focus, and editing for conciseness and clarity. These skills are certainly important, but the broadband mechanism is one of networked collaboration and media intensity, and the newer skills of writing modular and reusable content, integrating text and graphics, and building procedural systems rather than simply describing procedural steps are left underdeveloped. Even cutting-edge research in technical communication classrooms still tends to focus on print media, and strategies such as single-sourcing and metadata manipulation, while undoubtedly useful, are taught mainly in relation to print. Not surprisingly, problems emerge when technical communicators are presented with more dynamic and specific forms of media production and distribution (e.g., interactive video training materials, dynamic Web sites, and tools for virtual or augmented reality) in which familiar measures no longer apply or are altered significantly in form or function.

This focus on general or traditional writing, coupled with the rapid evolution of other technologies and ideas in industry, often complicates academic and industrial collaboration. With such a fundamental difference in perspective such as that toward production processes, it is difficult to find a means for building an intersubjective space (i.e., a common ground) between the two communities. One would hope that the research produced by academics would help drive the latest innovations in technical communication and documentation practices, but the two communities are much less likely to interact than one might think (Grove, 2005; Kress & Van Leeuwen, 2001; Mirel & Spilka, 2002), and academic faculty and industry professionals rarely publish *any* articles together (Palmeri & Tuten, 2005), let alone articles that might lead to ideas for resolving these issues. As Laurel Grove (2005) explains, "Practitioners and academics often fail to recognize their common interests—practitioners see research as irrelevant; academics see practice as lacking any theoretical basis" (p. 237).

Differences in access to resources further complicate the division between academic and commercial domains. Broadband technologies enable the use of video conferencing and real-time, distributed simulations of complex mechanical and procedural processes. Such technologies remove the requirement that team members and employees be co-located and encourage the spread of industrial tasks from a central location to a more decentralized and cost-efficient model of operation. As Sosa, Eppinger, Pich, McKendrick, and Stout (2002) note, various stages of industrial processes often take place in different areas of the world; large corporations commonly design hardware in one location, design software in another, and begin the manufacturing process in yet another. In an academic setting, it is much more likely for students to encounter all aspects of their training in a single location. On the positive side, internships, distributed learning, and Web-based coursework are beginning to offer universities some of these same types of cost and efficiency benefits enjoyed by industry.

BROADBAND MEDIA AS AN ACADEMIC CONSIDERATION

Given this rather polarized arrangement of academic and industrial goals and methodologies, it seems that technical communication pedagogy will need to continue to evolve in order to respond to the increasing prevalence of multimodal materials and broadband business practices. This argument presupposes, of course, that the goal of such curricula is to prepare the majority of graduates for industrial placement. Students headed for graduate school, however, will also benefit from a more robust treatment of the processes of designing, producing, and distributing products.

The intrinsic properties of broadband media and the ways in which these multimodal media are produced and distributed to an audience are important topics for the technical communication classroom. There are some notable differences between how information can be packaged and transported in a broadband environment and how it can be moved in more limited network environments. These differences range from the technological (e.g., more bandwidth suggests greater opportunity for interactive and immersive multimodality) to the rhetorical (e.g., user agency is enhanced, and reformulations of ethos are based on community-driven support forums and other feedback-based documentation initiatives). Academic study of these issues can offer insight, especially regarding the relation between bandwidth and Internet connection speed to technical communication design and production standards.

This task may be less daunting than it seems, since many properties of broadband media are refashioned from traditional media. Hypertext, for example, draws on and reconfigures earlier technologies such as alphabetic media, the telegraph, video, film, and animation (Bolter & Grusin, 2000). Even so, such juxtapositions and reapplications of older forms in newer interfaces inevitably create new rhetorical contexts, as Adrienne Lamberti discusses in her chapter in this collection. For instance, we might consider the notion of high-speed kairos for broadband technologies. In a broadband environment, virtual time can be shifted and reorganized using creative video editing techniques such as the shuffling of keyframes (the beginning and ending moments of a transition between video frames) in an Adobe Flash-based interactive tutorial. Space also can be altered through the creation of simulated three-dimensional training environments, allowing a viewer to perceive depth and texture through creative applications and manipulations of rendered art, wireframes (the layout of a visual interface's basic components), shadow, and lighting. The question from a kairotic perspective, then, might be which changes in the rhetorical act occur when the medium shifts from a linear to a random timeline, or from a two-dimensional to a three-dimensional environment?

This type of academic question can apply to other fundamental topics of rhetoric, including decorum and audience, as they relate to industrial contexts. For instance, how will a typical end-user feel about being able to rotate and to expand a vacuum cleaner in three dimensions in order to find a belt that needs replacing, instead of reading about this process in a bulleted list? Is such interactivity and fidelity even necessary? What type of empirical evidence might be gathered to support a conclusion on this point? What is more, the professional research of faculty members can be offered to students in order to demonstrate the value of both theoretical and applied academic work, as well as the benefits offered by the modest pace of academic inquiry.

The Broadband Age presents a unique vantage point from which to explore the implications of modern technical communication practice from an academic perspective. As another example, the design of products used for simulation training purposes requires more than the integration of multimedia and software development knowledge. A holistic view of product design, much like the principles and practices of systems engineering, are necessary if a product is to be useful as a training tool or as a job aid. In providing a framework for a systems engineering curriculum, Aaron Shenhar (1994) advocates a holistic view of product design that is "the design of the whole as distinguished from the design of the parts" (p. 328). A "holistic view," Shenhar argues, "is multidisciplinary in nature, rather than disciplinary or interdisciplinary" (p. 328). To be sure, as a new-media production, simulation demands application and tool knowledge, but, as Shenhar writes, "Complex systems can only be created through the combined efforts of many people" (p. 328). In the workplace, technical communicators will more than likely be an integral part of a product team. In the case of a simulated training product, training systems analysts typically conduct a needs assessment in the discourse stratum to identify the requirements of training and systems. Instructional systems designers develop the instructor tasks and functions and work with multimedia engineers to select the appropriate media for instruction and delivery. Concurrently, cognitive psychologists and learning theorists focus on students and learning objectives; both might participate in curricular design and storyboarding, as might graphic designers and multimedia specialists. A move into the production stratum occurs as software and electrical engineers design the system architecture and as mechanical engineers design the physical workspace. Lastly, during both production and distribution, human factors engineering ensures that ergonomic concerns are met; specialists in this field also participate in human-computer interaction analysis, usability studies, and system testing.

As the examples above illustrate, to better align rhetorical skills with industry needs, a more precise understanding of what happens as students move from academic training into the professional community is necessary. We next will look closely at technical communication in the industrial professions our students may enter. This analysis is derived from more than 28 years of experience working with military technology and technical documentation.

INDUSTRIAL PERSPECTIVES

Broadband technologies have had an especially powerful influence in the defense industry. Because of the unique situation in which technical communication materials must be used in this field, it is comparatively easy to study technical communication as it relates to a larger system design process using broadband media. In this section, we focus on industrial perspectives and the deliverable production cycle to identify different types of gaps that emerge when newly educated technical communicators enter the workforce. This focus will set the stage for the revised curriculum we suggest in the final part of the chapter.

Our analysis of deliverables here is focused on one particular type of document: the Interactive Electronic Technical Manual (IETM). We exemplify the disconnect between academic and industrial knowledge by illustrating how this type of deliverable has been held back by traditional conceptualizations of technical communication.

Paradigm Gap

We might begin by asking this question: Where does technical communication fit within the complex scenario of system design? What seems obvious is that the holistic and multidisciplinary approach advocated for system design should be extended to include interactive and intelligent technical documentation. But, surprisingly, technical communicators often view their documentation projects, whether new media designs or traditional linear documents, as distinct and separate products from the product being developed. These documents are not considered a crucial part of the whole system design, but rather a small and often insignificant part. And it is here within the technical publication groups that the differences in skill level created by differences in broadband adoption are most obvious.

Broadband-enabled technologies have exposed skill deficiencies within groups *of* and, ultimately, *in* individual technical communicators. An assessment of the design of an IETM shows just how substantial this skills gap can be.

IETMs on one level are nothing more than reference materials, but on another level are artificially intelligent applications meant to improve maintenance time and accuracy. The IETM concept model is focused on machine-to-machine interactivity, with human intervention when and if extra guidance is necessary. An IETM residing on a laptop computer is connected to a sensor embedded in another product, and some type of electronic control-unit interface allows a dialog to commence between the two computers involved. Depending on the outcome, the user will be instructed on a task to be completed, if any. For example, soldiers use IETMs to troubleshoot and to maintain their equipment in lieu of paper technical manuals.

Although technical documentation can be both interactive and multimodal, even when the documentation is tagged in specialized metadata markup languages such as Standard Generalized Markup Language or XML and embedded with media, IETM documentation usually represents the "look and feel" of a traditional manual. This reflects the fact that technical communicators are trained to remain faithful to the idea of traditional print writing techniques, while other industry professionals tend to be trained to view writing as secondary to their primary goals and tend not to offer much useful criticism regarding the selected content delivery system. As a result, technical communicators may use models more appropriate to print media than to broadband media, and these models, however outdated, continue to survive unchallenged. It should be clear that broadband technologies are capable of using high-bandwidth networks at full extent. Why is it, then, that the average IETM is no more than repurposed print enabled by low-speed broadband multimedia technologies?

Expert system design, along with developing queries, if-then-else prompts, scripts, and the intelligent user interface, is the result of the interplay of group members whose skills are diverse, such as cognitive psychologists, human engineers, instructional designers, and information technology professionals. When so many of the materials and exercises students encounter in the classroom are printcentric, it is no wonder that industrial employees do not readily understand the potentials of the increased bandwidth and speed provided by faster networks. Those who create IETMs could contribute significantly to designing such manuals if they followed a more creative and inventive style of multimodal communication—one that might be used to refashion a broadband medium more appropriate to the environment in which it is used.

Group Dynamics Gap

Group dynamics may also provide some insight into printcentric practices, for technical publication groups can be territorial. This can be another abrupt change from the academic classroom, where individuals are encouraged to work together. For the most part, this process is mediated by a central figure (e.g., the professor) who is likely to encourage democratic behaviors and respectful and open conversation within and across the groups. Often staffed and organized according to group member's skills, classroom groups tend to reinforce the craftsman model (Albers, 2003) of technical documentation development, with individual students contributing different sections of a group document and then a collective review of its contents. Although groups are *somewhat* heterogeneous, students in advanced technical communication courses are mostly strong writers and have similar basic skills and competencies gained from prerequisite courses.

On the other hand, in a traditional defense industry training organization, for instance, technical publication groups are *highly* heterogeneous. They generally consist of employees with specialized skills (unless in a truly small business), all of which contribute to the discourse, design, production, and distribution of the manual: Subject-matter experts with relevant experience, skilled writers and editors, word processors and end-product publishers, and technical illustrators. In the flight-simulation industry, people are employed based on individual subject-matter expertise and, as Katherine Staples (1999) notes, are "probably males, perhaps ex-military or a former technician, of middle age, and probably a long-term employee of a single company" (p. 156). These technical communicators are hired because they were once users of related equipment themselves and, as such, have worked intimately with similar systems and their supporting technical documentation. Changing organizational practice or attempting to revise existing technological practices (as suggested in the prior section) can be quite difficult for new employees because they have comparatively little experience with organizational or multidisciplinary culture.

Creative Skill Gaps

The increased capacities of broadband technology and its ability to showcase more media-intense forms of documentation such as animation and three-dimensional video games has elevated artistic skills in industry and surfaced related fissures between academic training and industrial procedures. Again, this is a case in which writing is stressed in the classroom at the expense of covering other potentially necessary skills. Currently, most industrial teams in the defense industry have other members of the technical publications group provide illustrations and graphics expertise.

While subject-matter experts and writers once sketched concept drawings for the technical illustrator in order to suggest a 3-D perspective of the product and its components, broadband technologies have enabled both technical illustrators and technical communicators to share design space. Technical communicators often are asked to use engineering drawings as the basis of a figure or graphic and work with the drafter to modify drawings or convert them to a format that can be imported easily into the desktop publishing application and modified by the technical communicators themselves. This is one example of an opportunity whereby technical communicators can position themselves as knowledge workers within the organization, yet it is unlikely that individuals will demonstrate this type of initiative or creativity when it has not been stressed as part of their academic training.

Summary and Analysis

To summarize, in the traditional workflow model of technical documentation development, those in the technical publications group not only are disconnected from the product development team in terms of space, but also may be segregated by skill within the group. Concerned with supplying the appropriate content, subject-matter experts interact with the product-development team but typically have little interest in the writing and editing phases or in the production and distribution phases of development. Writers and editors rely on input from the subject-matter experts but are not really concerned with how that knowledge is created and probably do not understand how their writing skills can enhance overall product design. Technical illustrators rely on input from writers and perhaps from subject-matter experts but have very little to do with the overall production of documentation. Concerned mainly with producing the final product and distributing it to the customer, word processors and lower-level tool users express little interest in the collection of the subject matter or in transforming it into a viable text. Some of these problems stem from the ways in which employees were trained in the classroom to think about their disciplines.

Within any organization, and especially in the engineering or technical industries, team members with a multidisciplinary point of view and with a holistic understanding of the product are much more highly regarded than those who work in isolated territorial environments. As Roger Grice (1994) points out, these individuals "can bring a broadened enriched perspective to the product being developed and produce something that may be more widely accepted in the marketplace" (p. 389). Such team members are considered technically literate, not because they have technical educations, but because they understand the team dynamic and how this dynamic works to achieve a marketable and usable product. They are proactive contributors, they consider the contributions of other disciplines crucial inputs into the design of technical documentation, and they view technical documentation as an integral part of the system, not as a stand-alone product or as a by-product. Such a team member profile is commensurate with the technological, economic, and communicative developments afforded by the Broadband Age.

We are not suggesting that valuable technical communicators must have an engineering or technical background, nor do we mean to imply that they should become subject-matter experts in the topics being written about. We do suggest, however, that technical communication curricula begin to focus on a diversity of skills and disciplines, much like that in systems engineering, whose "wide applications in various industries prove its effectiveness as a means of integrating distinct disciplines and technologies into an overall complicated purpose" (Shenhar, 1994, p. 327). Marjorie Davis (2001) adds, "Unless technical communicators want to remain in a servant role, [they] must become more than tool jockeys [and] complete the evolution from craftsperson to professional" (p. 139). Familiarity with a diversity of skills and disciplines can liberate technical communicators from their perceived role as "support" and elevate their status within the organization, as experts in both communication and product design. In our concluding section, we offer ideas for initiating steps in this direction.

A PROPOSED CURRICULUM FOR TECHNICAL COMMUNICATION

This final section of our chapter presents strategies for strengthening technical communication pedagogy for the Broadband Age. We also consider applying academic techniques to improve industrial practices. This holistic approach means improving and incorporating more interactive and tools-driven design techniques for the student during academic training and utilizing more inclusive and multimodal theories for the industry practitioner in the development and deployment of broadband documentation products. Such a model borrows from both communities, for the academic world is ideally suited for the advancement and pursuit of knowledge while industrial models leverage resources that lead to improvements in efficiency and innovative practical models for facilitating technological workflows. By considering the standpoints and practices that have led to divergences in these two communities' understanding of collaboration, writing, and technology, we can initiate a conversation between experts in both domains regarding how to narrow methodological and epistemological fissures.

Before outlining our plan, we wish to acknowledge that this model evolves out of our own focus on military training, maintenance, and simulation documentation. We feel that many of these discipline-specific ideas are worth considering at a more general level for two reasons. First, new media products in this field are perhaps uniquely complex, demanding a very high degree of innovation in terms of text and technology, all of which *future* broadband forms of documentation almost surely will need to possess. Second, there is a long and well-studied shared history between military and academic applications of technical communication. This rich history demonstrates crucial interactions between media and rhetoric that, we believe, are useful for shaping a unified model of pedagogy. Before moving to our curricular suggestions, we will briefly examine a theoretical framework useful for merging key ideas from academia and industry.

Reconciling Academic and Industrial Objectives

With the functional and rhetorical dilemmas from the earlier parts of this chapter in mind, we turn to a theoretical model that may address some of the divergent paradigms that can frustrate academic and industrial collaborations in the Broadband Age. The academy and industry are obviously different institutions with different goals and practices, and we do not mean to suggest that they should have the same vision, nor that they should share the same goals; indeed, we consider their differences to constitute an advantage to society. Rather, we suggest that a shared understanding of production is useful and can help align educational goals with the professional skills required by employers. We suggest specifically that a theoretical model can help identify divergences between the production process as a student understands it when graduating, and the production process as it exists in industry.

This model of multimodal communication, developed by Kress and Van Leeuwen (2001) and mentioned at the beginning of our chapter, is especially appropriate to our task for three primary reasons. First, it is *practice based*. Second, it is *multimodal*, meaning that it can account for traditional text as well as for the juxtaposition of such text with audio and video materials. Third, it does not abandon the socially constructed influences of meaning making, but rather incorporates these elements as integral to its theoretical foundations. Although this model was not constructed initially to deal with interactive, high-speed media, it is flexible enough to be extended in that direction, toward the rich media applications so often dependent on broadband technologies. In short, we offer one interpretation of Kress and Van Leeuwen, from the perspective of technical and professional communication.

As discussed earlier, Kress and Van Leeuwen's multimodal theory of communication is built on discourse, design, production, and distribution strata, or domains of practice. Discourse is found in those social spaces where specialized knowledge is shared and exchanged (p. 4). In our discussion thus far, we have suggested that discursive knowledge gaps exist between academic and industrial communities of technical communicators, for a variety of reasons.

Design is described by Kress and Van Leeuwen as standing "midway between content and expression" (p. 5). In this sense, design can be considered the articulation of an idea in an abstract form. The design phase is important because it serves as an intermediate stage in which methodological or technical wrinkles associated with a product can be ironed out. Design products may take on tangible shapes as well, in forms such as design blueprints for a given architectural structure (though not necessarily the final blueprints), in written notes for a given meeting, in notes taken during a meeting, or in other intermediary mental or articulated states. Even the deliverables produced through most technical communication courses—documents such as instruction manuals, project plans, and even software documentation files—are more characteristic of design than of production, for these so-called final projects are imperfect and in a state of revision, with the final assessment of quality provided by the course instructor. Were a final project to be carried over from one course to another and handled by multiple groups with multiple agendas and foci for assessment, the type of transfer found in the production stratum would be reproduced more authentically.

When these abstract ideas or intermediate forms are moved into more permanent and architecturally sound states, they become productions (p. 6). Examples of productions are the construction of a building from a set of blueprints or, more generally, the use of material resources to commit abstract ideas to fully realized forms. In industry, these forms might be represented by final-release versions of documentation, by shipped copies of software manuals, by the launch of a new documentation Web site, or by the movement of an interactive electronic technical manual from Beta (pre-release) to non-Beta (release) status. In contrast to the design stratum, however, the production stratum involves no mediator standing between designer and audience—no instructor looms over the product and assesses whether the articulated form and its medium are appropriate for the intended audience or audiences. In the production model, errors are minimized and the product is shaped into its final form and prepared for distribution.

Final products are disseminated to an audience in the distribution stratum. While this may introduce new theoretical or philosophical considerations—Kress and Van Leeuwen note Walter Benjamin's (1977) work on the loss of the "aura" of authenticity when an object is removed from its original context and mechanically reproduced and relocated—this phase in industry typically is much more concerned with the practical and economic issues involved in providing an audience with access to production materials (p. 7). Digital media environments often integrate production and distribution strata, in that digitally mastered forms of artwork and documentation can simply be moved from one server to another in order to enable worldwide access to materials.

Although these four domains may be accessed by a student in a classroom or by an employee in an industrial setting, there usually are more opportunities for discourse and design to be discussed in an academic setting, where control of both design and synthesis of academic products can remain with an individual student. In industry, it is rare for an employee to follow a design from abstraction to concrete formulation; instead, a given employee may inherit a task after it has gone through its initial conceptualization and been vetted for production by management personnel or committees. Insofar as teachers cannot anticipate the position a specific student may end up securing upon graduation, it seems sensible to us to engage all four discourse and design strata in technical communication curricula, and especially to prepare students for high-speed production and distribution, processes they are likely to encounter in their workplaces.

In considering this scaffolding for multimodal synthesis, we can begin to develop a more precise theoretical understanding of the types of skills necessary in the Broadband Age. A preliminary attempt at such a task, which considers not only examples of existing formative products used by technical communicators but also the media in which those products are made evident, might include examples of activities such as (a) audience analysis, (b) project management, and (c) help documentation; along with their corresponding media, such as oral interviews or ethnographic studies, Gantt charts or project management plans; and HTML help documentation, compiled help documentation, or XML/docbook help documentation. Although these examples do include media and modes from several of Kress and Van Leeuwen's strata, including discourse, design, and production, they are oriented more firmly in the former two strata.

What is immediately evident from this assessment of both tasks and theory is that technical communication in broadband and networked industries increasingly is becoming a multimedia-enabled process; employees building interactive documentation systems will often work with two-dimensional and three-dimensional graphics, audio files, computer programming scripts, static and dynamic forms of text, and interactive video. Additionally, they will be expected to interface with hybrid groups composed of management, graphic designers, project managers and team leaders, subject-matter experts, and clients. While the selected "traditional modes" outlined in this discussion do much to address these actions and interactions, a more marketable professional portfolio may in fact involve synthesis and cultivation of the additional (and oftentimes multidisciplinary) skills we might define as "broadband skills."

Practical Suggestions

While we hope that this theoretical treatment is useful for conceptual alignment of paradigms between the academic and industrial perspectives identified in this chapter, we also need practical and immediate solutions to address gaps in skill and practice. An immediate step in the right direction would be to move from an isolated model of professional training to a more integrative and multidisciplinary model. Based on our analysis of current academic trends and technical publication groups working in the field of simulation and training, we urge technical communicators to immerse themselves in overall product design. They should not continue to work, as Grice (1994) describes it, "in an isolated, support group," but attempt to become "an integral part of the product-development team" (p. 389). According to Grice, "Those who can thrive in the environment will be able to make significant contributions to the product being developed; those who cannot, on the other hand, will make smaller contributions, if any, and may be perceived more as a liability than as an asset" (p. 389).

This is painfully true in the simulation and training industry. Technical communicators who are limited in terms of the technological skills they can bring to the table are considered a burden to the team, and their products are often considered archaic. And, as a result, interactive technical documentation with true interactive functionality often is given over to information technology experts for development, not to technical communicators. The Broadband Age presents technical communicators with an opportunity to reverse this trend, but this will not occur without a change in practice by both educators and managers.

What, then are some of the core competencies a technical communicator should develop for the Broadband Age? Corey Wick (2000) suggests that "technical communicators develop a more comprehensive understanding of the technologies that service knowledge management as well as business in general" (p. 527). Davis (2001) suggests that technical communicators should have a basic understanding of technological competencies, management, and teamwork skills. We believe all these skills are required for the future of technical communication and propose that technical communicators must also be versatile in a wide range of skills and disciplines.

Our assessment of the current state of affairs in both academics and industry reveals several overall factors leading to deficiencies in professional, entry-level skills for technical communicators entering the workforce immediately after completing their undergraduate educations:

- They do not have multidisciplinary backgrounds.
- Their focus is on the product as a separate entity rather than as part of a unified process.

- They continue to support and to be trained in the craftsman model of technical documentation, with minimal attention paid to more modular methodologies.
- The skills they do have are consistent with skills enabled by low-bandwidth technologies.
- They do not have the skills to take advantage of high-speed-enabled multimedia technologies.

To address these deficiencies, we suggest four practices that span both academic and industrial contexts and that we feel are important and relevant to the Broadband Age. First, we feel that both academic and industrial attention to the characteristic of computational interactivity is critical. An approach compatible with that of IT will be vital to the survival of technical communicators in broadband and networked economies. Such partnerships already are being formed with library sciences, for example, under the term "information literacy" (see D'Angelo & Maid, 2004). Understanding the various interactions between teams of humans and computers will be crucial for designing effective new types of documentation products situated in broadband digital space. As interactivity is what separates "new" media from "old" media (Murray, 1997; Zappen, 2005), it is something that needs to be continually stressed and explored in both education and industry.

Second, we feel that a multidisciplinary education is essential. For technical communicators to break out of the craftsman model of technical documentation development, and for students to understand how other team members can provide valuable input into documentation development, technical communication curricula should include courses in human factors, instructional systems design, IT, digital media, and cognitive psychology. These disciplines are involved in overall system design, and these specific disciplines are useful for a variety of industrial tasks, not just for those associated with military simulation and training. It is not practical or fair for us to ask the typical technical communication professor to learn the wide array of skills necessary to educate students in the many skills they will need for working in modern industry. It is fair, however, to ask program administrators to recognize the value of different skills and how they fit into the overall matrix of technical communication pedagogy, and to arrange students' coursework around this multidisciplinary structure. Recognizing the inevitable political battles and turf wars that may occur here, we conclude that this will perhaps be the most difficult change to effect in academe.

Third, we believe that collaborative internships are vital for establishing a linkage between academic skills and practical, on-the-job needs. Internships in technical communication provide valuable insight and experience for discipline-specific and industry-specific types of problem solving. Technical communicators who intern at large and small businesses will ideally have the opportunity to see firsthand where their contributions have the greatest impact. They will also learn if their core competencies and technical skills are adequate for workplace success. Moving curricula outside of the classroom and into the field has interesting implications for technical communication students hoping to catch a glimpse of their future professional lives before actually beginning them. Such a strategy has proven successful in rhetoric curricula (McGill, 2003); similar benefits likely can be found in arrangements whereby technical communication students conduct ethnographic research on industrial teams.

Lastly, we feel that it is important for technical communicators to understand and to appreciate the business-oriented mechanisms driving industrial practice. Management and leadership courses may be useful for this goal. Curricula that include program-management studies—cost strategies, scheduling, risk-mitigation techniques, resource management, team building, and customer relations—at the product level will help technical communicators understand how documentation fits into overall project development and which factors ultimately influence its success. Knowledge of these practices can also be helpful for understanding how different teams and groups work together with common goals in mind.

CONCLUSION

In this chapter, we explored the ways in which digital divergence is affecting the simulation and training industry by giving rise to a gap in understanding and practice between communicators who are equipped to deal with broadband technologies and communicators who are not. We offered examples from the defense industry to highlight the key trends we believe are affecting the field of technical communication and will be affecting it for some time. Although certain differences between academic theory and practice and industrial theory and practice are longstanding, we have attempted to demonstrate in this discussion how these differences are being exacerbated as a result of the challenges that broadband technology poses to both producers and consumers of media.

Using Kress and Van Leeuwen's multimodal theory of communication, we examined the function of technical communication from an academic perspective focused primarily on discourse and design, and from an industrial perspective focused primarily on production and distribution. Later, we revisited this model to examine the shifts in thinking that may be necessary to better align practices in academe and industry. Finally, we identified several problematic factors that may disadvantage future technical communicators and offered preliminary suggestions that might prove useful remedies. It is our hope and expectation that researchers will continue to investigate methods for empowering technical communicators in IT-laden environments and to encourage interdisciplinary and collaborative initiatives incorporating the benefits of broadband technologies.

REFERENCES

Albers, M. J. (2003). Single sourcing and the technical communication career path. *Technical Communication, 50*(3), 335–344.

Arbori, A., & Ordanini, A. (2006). Broadband divide among SMEs. *International Small Business Journal, 24*(1), 83–99.

Bolter, J. D., & Grusin, R. (2000). *Remediation: Understanding new media*. Cambridge, MA: MIT Press.

D'Angelo, B., & Maid, B. M. (2004). Moving beyond definitions: Implementing information literacy across the curriculum. *The Journal of Academic Librarianship, 30*(3), 212–217.

Davis, M. T. (2001). Shaping the future of our profession. *Technical Communication, 48*(2), 139–144.

Grice, R. A. (1994). *Technical communicators in product development teams: What can we really contribute? Scaling new heights in technical communication.* IPCC '94 proceedings (pp. 389–393). Alberta, Canada: IEEE Professional Communication Society.

Grove, L. K. (2005). Research reality check. *IEEE Transactions on Professional Communication, 48*(3), 237–241.

Harner, S., & Rich, A. (2005). Trends in undergraduate curriculum in scientific and technical communication programs. *Technical Communication, 52*(2), 209–220.

Kress, G., & Leeuwen, T. V. (2001). *Multimodal discourse: The modes and media of contemporary communication*. New York: Oxford University Press.

MacDonald, C. (2002). Theories of mind and "the commonsense view." *Mind & Language, 17*(5), 467–488.

McGill, K. (2003). Field study and the rhetorical curriculum. *Technical Communication Quarterly, 12*(3), 285–302.

Miller, C. (1979). A humanistic rationale for technical writing. *College English, 40*(6), 610–617.

Mirel, B., & Spilka, R. (2002). Introduction. In B. Mirel & R. Spilka (Eds.), *Reshaping technical communication: New directions and challenges* (pp. 1–6). Mahwah, NJ: Lawrence Erlbaum.

Murray, J. H. (1997). *Hamlet on the holodeck: The future of narrative in cyberspace*. New York: Free Press.

Palmeri, J., & Tuten, P. M. (2005). Dialogic negotiations: A reflective tale of collaboration across the academic-practitioner divide. *IEEE Transactions on Professional Communication, 48*(3), 313–323.

Rutter, R. (1991). History, rhetoric, and humanism. *Journal of Technical Writing and Communication, 21*(2), 133–153.

Scollon, R. (2003). The dialogist in a positivist world: Theory in the social sciences and the humanities at the end of the twentieth century. *Social Semiotics, 13*(1), 71–88.

Shenhar, A. (1994). Systems engineering management: A framework for the development of a multidisciplinary discipline. *IEEE Transactions on Systems, Man, and Cybernetics, 42*(2), 327–332.

Sosa, M. E., Eppinger, S. D., Pich, M., McKendrick, D. G., & Stout, S. K. (2002). Factors that influence technical communication in distributed product development: An empirical study in the telecommunications industry. *IEEE Transactions on Engineering Management, 49*(1), 45–58.

Staples, K. (1999). Technical communication from 1950–1998: Where are we now? *Technical Communication Quarterly, 8*(2), 153–164.

U.S. Department of Commerce. (2005). *A nation online: Entering the broadband age.* Retrieved 30 November, 2005, from http://www.ntia.doc.gov/reports/anol/index.html

Wick, C. (2000). Knowledge management opportunities and leadership opportunities for technical communicators. *Technical Communication, 47*(4), 515–529.

Zappen, J. P. (2005). Digital rhetoric: Toward an integrated theory. *Technical Communication Quarterly, 14*(3), 319–325.

http://dx.doi.org/10.2190/CWDC11

CHAPTER 11

Sizing Up Single-Sourcing: Rhetorical Interventions for XML Documentation

Aimee Kendall Roundtree

OVERVIEW

This chapter discusses the difficulties often plaguing application of extensible markup language (XML), a popular technology for single-source digital documentation, which encodes text files with tags that distinguish between content types or categories. Common XML indexing practices often fail to reflect the best usability practices of technical communication. This is so because technical communicators on single-source project teams often do not know enough about XML to be able to recommend how to use such practices and because XML programmers in turn often do not know enough about best practices to implement them. This chapter introduces technical communicators to XML and underscores the importance of applying their knowledge of rhetoric, usability, and human factors to single-source documentation.

Content management systems, or technologies for creating, organizing, modifying, and removing data contained in a database or content archive, have made it easier for businesses to institute *single-source documentation*, or the creation of several deliverables from one such database or archive. The rush toward single-sourcing has obliged many technical communicators to familiarize themselves with new software and computer languages. This chapter introduces and discusses one popular technology for single source digital documentation: *Extensible mark-up language* (XML), which offers a means of encoding text files in order to distinguish among different content types or categories.

Extensible mark-up language is an indexing technology that allows authors to categorize data types within written content so that computers can search through content, pick and choose pertinent parts, and generate multiple combinations of content segments. In this way, one XML document with multiple levels, layers, and types of data can become the source for a range of deliverables. The mark-up language transforms a text document into its own database wherein the author can define chunks of content by function or type. By categorizing content, XML makes it easier for computers to conduct context-sensitive searches within documents and to reuse crucial segments (Rockley & Manning, 2000, pp. 1–2).

The advantages of XML have been discussed widely. Some researchers report that XML has made manuals, Web content, and other documents easier and less expensive to create. For example, pharmaceutical company GlaxoSmithKline uses an XML-based system to translate the company's container labels, advertisements, inserts, and other documents into a dozen languages (Rappaport, 2002, para. 3). Rather than write and approve hundreds of different Word documents containing the content for each of these deliverables, the company edits a master document and creates ancillary deliverables from subsets of information contained within it. Streamlining the editorial process can also help decrease production costs. Adam Jones reports on one company that over the course of three years was able to decrease source text changes and thus the number of translation updates (Jones, 2002, p. 3). J. D. Edwards reports on another company that saved $3.5 million by single-sourcing three documents in seven languages (Hackus & Hedlund, 2001, p. 2).

But widespread XML use in the technical communication field imposes a new demand for skills beyond writing proficiency (Rudder, 2002, 3.3 The Relationship . . . section). Perfectly logical XML code can and does produce audience-insensitive deliverables, and standard XML indexing practices do not always correspond to best usability practices as recognized by the technical writing community.

For single-source projects, technical writers typically serve as content editors and writers. But single-source project teams usually also include information designers (whose job it is to divide that content into reusable subcomponents to be flown into multiple deliverables) and information technologists or programmers (who code the XML) (Rockley, 2001, pp. 192–193). Information designers on single-source teams must have a mastery of writing skills and a familiarity with XML. They must understand the strengths, limitations, and capabilities of XML in order to know how best to structure, organize, subdivide, and categorize the complex and copious amounts of primary content and documents that single-source projects involve. Therefore, it often falls to professionals from user interface design, human-computer interaction, online instructional design, Web usability, and other computer-oriented fields to assume the role of information designers. Technical writers can facilitate the information designer's role—if not step into the role of information designers themselves—because technical writers have very strong writing skills. Technical writers are experts in chunking, paragraphing, subheading, and other essential writing skills. Unfortunately, there is often a gap between what technical writers know about XML versus what information designers and programmers know about it. The more technical writers know about XML, the more likely it will be that they will have the flexibility to apply best technical communication practices in writing and editing content for XML.

This chapter describes key XML features and explores how best practices can assist the technical communicator participating in single-source teamwork. Because XML enables closer collaborations among technical communicators, information designers and programmers, and experts in allied fields, but standard XML coding often disregards usability, readability, or other significant communicative concerns, this chapter explains how single-source teamwork can at times result in less than ideal documentation. Finally, the chapter examines and suggests strategies from usability research and the literature on *literate programming*—a reader-friendly approach to writing in computer languages—to help technical communicators compose rhetorically sensitive tagging schemes and navigate the stages of single-source workflow.

FUNDAMENTAL FLEXIBILITY OF XML SYNTAX

This section introduces key concepts for understanding XML in order to underscore how it can help technical communicators devise their own system of codes and descriptors. This flexibility will be elaborated on later in the chapter, as will rhetorical inventions for improving the user-sensitivity of XML.

Comparing XML to hypertext markup language (HTML) is a logical strategy for learning what XML is and what it does. Whereas in the description and creation of document formatting, HTML uses tags such as bold (<b></b>), italics (<i></I>), paragraph (<p></p>), and table (<table></table>), XML uses tags to describe different data types within a document (Craft, 2004, p. 2). Take, for example, a writer attempting to create a hypertext document about her family history. Were she to render the information in HTML, the tags would dictate the formatting of the document:

```
<html>
<head><title>Jones Family History</title></head>
<body>
<p><h1>Jones Family History</h1></p>
<p>Grandma: Irma Jones born in February 1915</p>
<p>Grandpa: Cleveland Jones born in July 1912</p>
</body>
</html>
```

Here, HTML tags separate data into paragraphs (<p></p>) and indicate a larger font for the heading (<h1></h1>) than for the rest of the document; a Web browser displays this formatting on screen.

In contrast, XML renders this information in terms of its content rather than its formatting:

```
<?xml version="1.0" encoding="ASCII">
<family_history>
  <member type="grandma">
```

```
    <name>Irma Jones</name>
    <birthday> February 1915</birthday>
  </member>
  <member type="grandpa">
    <name>Cleveland Jones</name>
    <birthday> July 1912</birthday>
  </member>
</family_history>
</xml>
```

In this example, XML tags indicate the type of content. XML contains tags for elements (such as <member> and <family_history>) that define broad categories of content. Some elements are subsumed beneath other elements. The element <family_history> contains the subelement <member> and thus is the *parent* of <member>. The element <member> is the parent of <name> and <birthday>. *Children* elements inherit the rules set for the parent. The content that tags define, such as "Irma Jones" in the preceding illustration, is called *character data*. The relation between elements and character data is, very loosely, like the relation between a mathematical variable and its value insofar as the connection between terms is arbitrary. Later, this chapter will describe how the freedom to name tags can be used more or less precisely to describe chunks of content. Creating tags that retain as many content clues as possible is a skill that technical communicators can bring to single-sourcing projects.

Unlike HTML, which is based on a set of universal tags defined by those who initially programmed the language, XML allows the writer to write his or her own human-readable tags, provided they follow the rules of XML syntax. The tags <member>, <name>, and <birthday> were my own invention. I could have made up other tags to indicate the meaning of each piece of data so long as the tags abided by XML rules, which follow the same general naming standards for most other mark-up and programming languages: Use an underscore, period, or hyphen instead of a space; avoid using special characters; respect case sensitivity; open parent elements before opening children; close children elements before closing parents.

XML writers also can invent *attributes,* or information that further modifies and describes the elements. The relation between elements and attributes parallels the relation between nouns and adjectives; attributes describe the extent, breadth, width and other features of the element. In XML, both tags and attributes have values. Were I to code the adjective-noun relation "soccer ball" in XML, I might use a tag called "object" to indicate the word *ball*, and an attribute called "type" to indicate the word *soccer*: <object type="soccer">ball</object>. Likewise, in the above example, the <member> tag contains an attribute called "type" that designates which family member is listed. The tags <member> and <name> separate one family member (and his or her name) from the other. The attribute "type" indicates the specific family member and name. For example, in the prior XML code, "Cleveland Jones" is the value of the member tag (a parent tag) and the name tag (a child tag). The attribute "type" contains the value "grandpa," which indicates which family member Cleveland is. That is, the values of the attributes assigned to describe each family member are "grandma" and "grandpa."

The process of naming and categorizing these parents, children, elements, and attributes lends XML strength for single-sourcing. The extent to which a single document can be used for multiple purposes and in multiple contexts depends on how detailed and predictive the XML coder designs the range of elements and attributes. Tags and attributes must be written in a way that anticipates various purposes, genres, and media for which bits and pieces of the document's content will be compiled and combined. For example, to create a separate list of all the grandparents, a single-sourcing team would create a new document that calls forth all types that are "grandma" and "grandpa"; in turn, a separate document listing all parents in the family would call on all types that were "grandma," "grandpa," "mom," and "dad."

What writers name elements and attributes also matters for efficient information architecture and searching. For example, programmers often build search engines to look for data at the level of the element tag, so the tag names become the key terms pointing to a specific type of content. From an indexer's perspective, technical communicators can make search engines more useful by including a wide and fixed set of indexical elements (Applen, 2001, p. 204). But if single-sourcing teams subdivide the information excessively, they run the risk of overcomplicating the indexing and making it more difficult for search engines to find key data. Lori Brown and Christian Walters (2004) put it succinctly: "The finer your grains, the more reusability you have, but the harder [information is] to find. More coarse [sic] grains are easy to locate but more difficult to reuse" (p. 4).

Writers of XML must break information down into independent chunks, identify their purposes, and classify them. Fortunately, it has always been the job of the technical communicator to find and to exploit information patterns within a text: We use parallelism to reorganize and to rewrite lists, and formatting to cue readers about different levels of information. Our habit of striving for clarity and conciseness can guide us in the creation of XML.

OBJECT LESSONS IN STANDARD INDEXING PRACTICES

Problems arise when the information designers and programmers who code XML overlook the dictates of audience and genre. Single-sourcing is often interpreted as a means by which one type of content suffices for all audiences and occasions, rather than as a means by which a specific content is adapted to different purposes and audience needs.

Despite the inherent compatibility between XML basics and the tenets of technical communication—categorization, simplification, and adaptability—the products created from single-source documents often fall short of ideal, in part because of the division of labor required. The source document must produce most (if not all) possible variations of deliverables. Therefore, XML code defining that source document often resorts to the lowest common denominator among genres—the more universal the category (the more frequently it recurs in different deliverables), the more likely a team will be to create a tag for it. Such universality can result in oversimplified indexing schemes. The following section discusses two instances of single-sourced documents in which content indexed with XML tags is flown into deliverables from different media and genres. Although these examples could be considered successful single-sourcing final products, they fail to meet usability standards in light of the genres and media used. Each instance illustrates how XML coding that is oversimplified or uninformed by the best practices of technical communication can result in user-unfriendly documents, and each might benefit from input from rhetorically attuned technical communicators.

iCalibre Example

Even the best coding can yield inadequate deliverables. Take, for example, this sample instructions template designed by iCalibre, a documentation company in Bangalore, India. The company posted this sample online as a demonstration of its typical methodology for single-sourcing documents:

```
<?xml version="1.0" encoding="utf-8"?>
<!— (C) Copyright IBM Corporation 2001, 2005. All Rights Reserved.*—>
<?xml-stylesheet href="../../xsl/ditaview_shell.xsl" type="text/xsl"?>
<!DOCTYPE task PUBLIC “-//OASIS//DTD DITA Task//EN” “../../dtd/task.dtd”>
<?altova_sps C:\Program Files\Altova\Authentic2005\sps\Template\dita\dita.sps?>
<task id="install">
<title>Insert Title Here<title>
<taskbody>
  <context></context>
  <steps>
    <step>
      <cmd>tbd</cmd>
      <info><note>Note inside a step instruction.</note>Some info</info>
    </step>
    <step><cmd>tbd</cmd></step>
    <step><cmd>tbd</cmd></step>
  </steps>
</taskbody>
</task>
```

(Vasudev & Patel, 2006)

The XML scheme accounts for many sublevels of data. Writers can

1. use open-and-close <cmd> tags to write steps of instructions,
2. introduce the context of the instructions with the open-and-close <context> tags, and
3. write feedback statements amplifying a step through the use of open-and-close <info> and <note> tags.

This XML scheme makes intuitive sense because it follows the “theme and rheme” structure of typical instructions. M. A. K. Halliday and Christian Matthiessen (1994) analyzed the structure of English sentences and highlighted the distinction between *theme,* that is, the starting point or “point of departure” for a sentence, where many writers place “given” details, and *rheme*, or the main point or destination of the sentence, where many writers place new details (1994, pp. 66, 79). For example, I have underlined the theme of the following sentence and left the rheme without text decoration:

> In the File menu, select Preferences.

The theme’s position at the front of the sentence conventionally lends it importance and emphasis; as a result, awkward sentences can result from misplaced and mistaken theme and rheme attributions. In the iCalibre sample above, which describes how to XML code a set of instructions, each open-and-close <cmd> tag would accept its own theme-rheme sentence describing a step in the instructions.

Likewise, iCalibre posted a template for a typical manual:

```
< <?xml version="1.0" encoding="utf-8"?>
<!-- (C) Copyright IBM Corporation 2001, 2005. All Rights Reserved.
*-->
<?xml-stylesheet href="../../xsl/ditaview_shell.xsl" type="text/xsl"?>
<!DOCTYPE concept PUBLIC “-//OASIS//DTD DITA Concept//EN”
“../../dtd/concept.dtd”>
<?altova_sps C:\Program Files\Altova\Authentic2005\sps\Template\dita\dita.sps?>
<concept id="need">
  <title>Why Should You Read This Book?</title>
  <conbody>
    <section><p>
```

With the world preparing for widespread acceptance and use of Open Source operating systems, [iCalibre] thought it exciting to prepare ourselves for open source technical writing tools. We asked ourselves whether serious technical writing can be supported by the tools available today.

```
  </p></section>
  </conbody>
</concept>
```

(Vasudev & Patel, 2006)

Here, four sublevels of data give writers a degree of flexibility. They can

1. use open-and-close <concept> tags to write multiple paragraphs on a theme or for making a specific point,

2. distinguish themes or points from one another by using attributes of the tags,
3. introduce different subparts of the manual using open-and-close <section> tags, and
4. designate headings with open-and-close <title> tags.

Ultimately, documents made with the software FrameMaker can import and use these tags to create online help software and Web sites. Figures 1 and 2 provide illustrations of documents that contain the same content in different formats.

Figure 1 illustrates how generic content from the XML master file is flown into what appears to be a standard help software that might be included on a CD for software installation. The software was designed to display the names of the XML pages in the navigation bar (to the left in Figure 1), the <title> content in bold, and the <conbody> content in normal font style. In contrast, in Figure 2 the very same tags produce what appears to be a standard HTML-based Web page. No alterations have been made to content. The XML code has created two entirely different deliverables in entirely different media using the same content.

Brio Example

Even using software such as FrameMaker (with which writers can look at a "what you see is what you get" [WYSIWYG] screen while generating XML) entails many of the same modular writing issues (Sapienza, 2004, p. 400). Notice how in

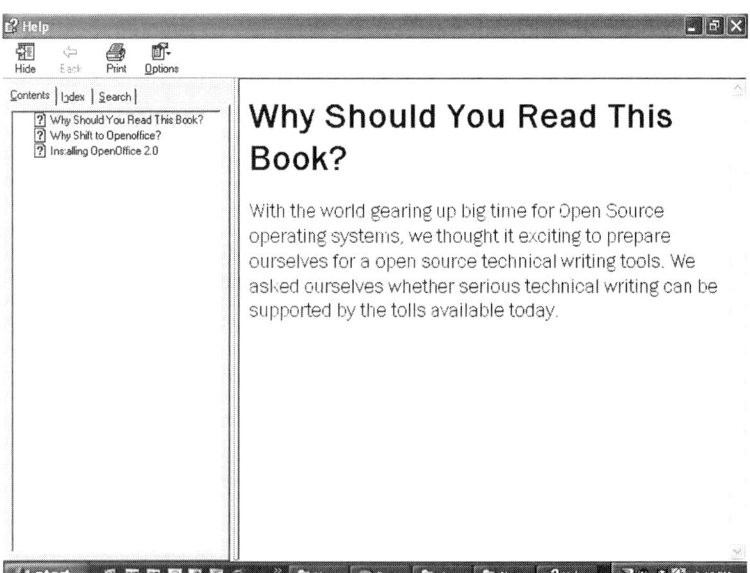

Figure 1. The iCalibre software manual. The XML-tagged content in applet format (Source: Vasudev & Patel, 2006).

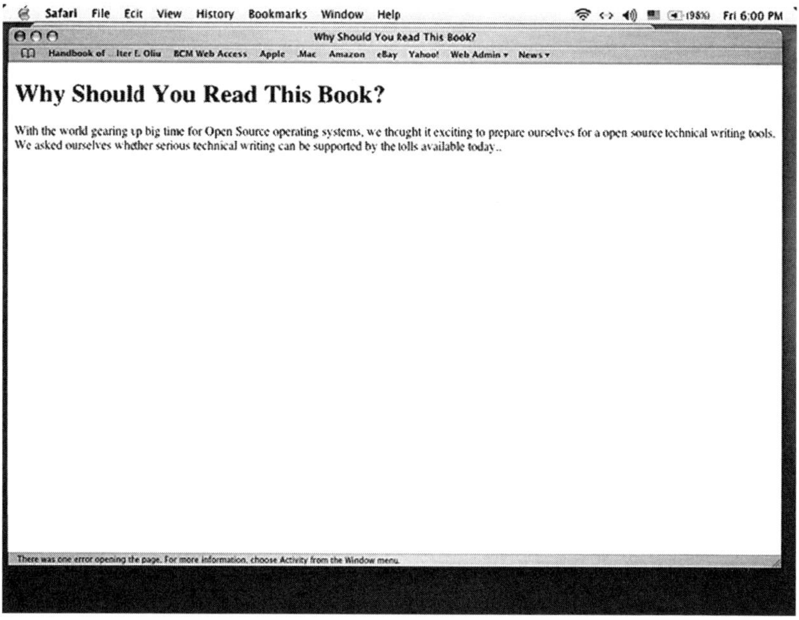

Figure 2. The iCalibre online manual. The same XML-tagged content from Figure 1, now in Web page format (Source: Vasudev & Patel, 2006).

Figures 3–5 technical communicators at Knopf Online—a technical communications service provider in San Francisco, California—revised the FrameMaker templates originally created by Brio to repackage the same content into PDF, HTML, and as online help software; the company then posted these documents as online samples of its work.

The "Charting Basics" page (Figure 3) is part of a manual that teaches Brio customers about a suite of business intelligence tools to aid in analysis and reporting. The chart feature is just one of many that comprise this suite. As with most PDF documents, this manual comes with a table of contents, index, and other features borrowed from print manuals. Through XML, the very same content can be imported into a Web page, as shown in Figure 4.

With regard to the Web site, Knopf coders do show some rhetorical sensitivity to genre; on the Web page, they include only the first subdivision of content—"Charting Basics"—from the corresponding PDF page. Otherwise, the content is identical, as it is when imported into the online help software applet (Figure 5).

These five samples illustrate some of the problems that arise when users of XML neglect issues of audience and media. In both the iCalibre and Knopf examples, content changes minimally (if at all) from deliverable to deliverable. One might argue that these authors have assumed that users of the three types of documentation

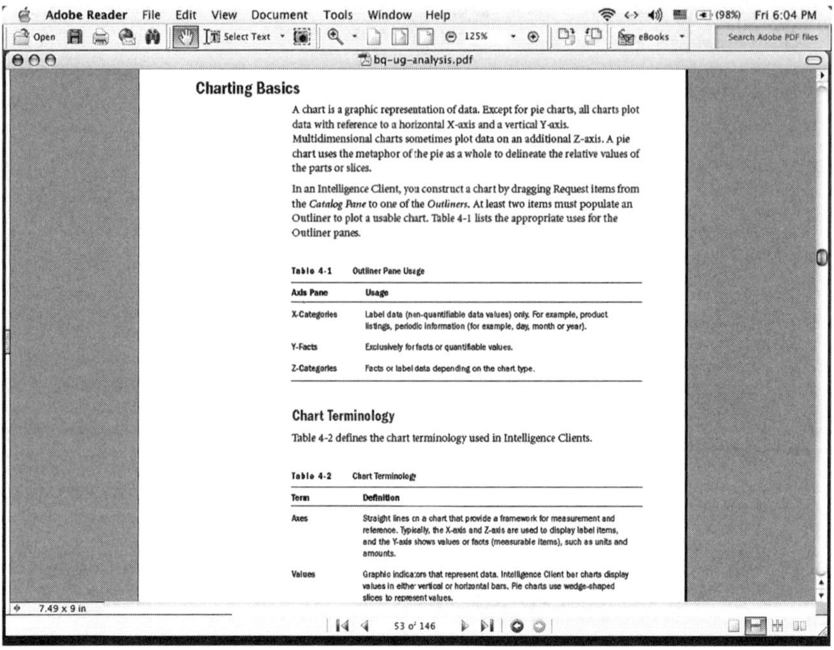

Figure 3. The Brio software manual, in PDF format (Source: Knopf Online, 2004).

require approximately or precisely the same information, that different media do not have different content requirements or expectations, and that usability will not be affected when content is unchanged across media. These assumptions contradict the best practices of technical communication, in which audience expectations are treated as context dependent, and content and format are tailored to occasion, genre, and medium.

Unless technical communicators with rhetorical expertise continue to supervise document production, XML will produce deliverables inadequately targeted to audience needs—for instance, manuals too complex for users, or Web pages too wordy for surfers. Further, certain theories of genre envision "text type" as more than a manifestation of formatting and syntax, but also as a marker for sociological norms and habits (Bazerman, 1994; Yates & Orlikowski, 2002). If it is true that genres manifest social and cultural expectations and practices, then XML coding that merely categorizes semantic and syntactic patterns and ignores pragmatic concerns will produce unsuccessful texts.

Many studies illustrate how different audiences require differently tailored details and different depths of definitions and explanations (Mirel, 1991; Scott, 2004). These audiences would have varying complaints about and expectations for the content of the Brio documentation, which Figures 3–5 illustrate remains unchanged from one medium to the next. Users who prefer online help, for instance,

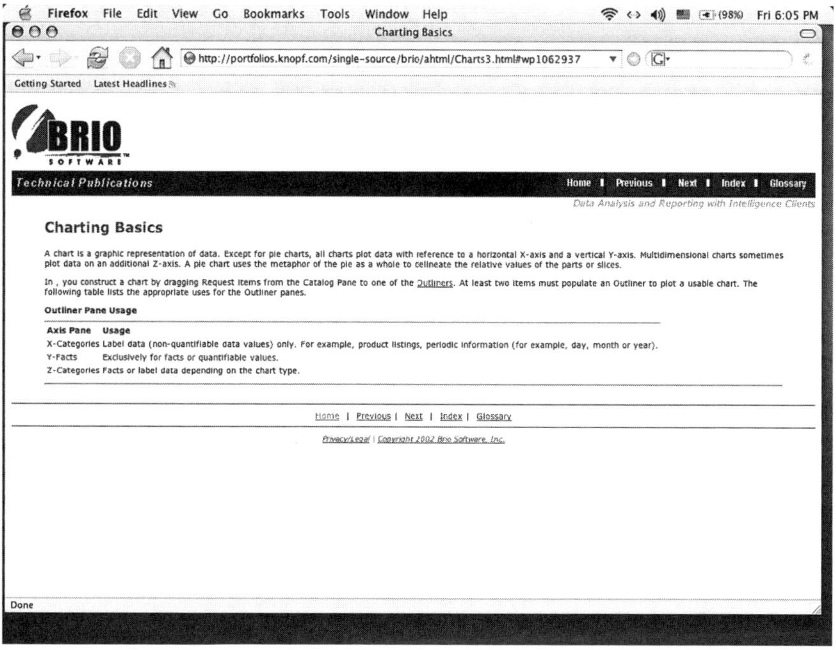

Figure 4. The Brio Web page, in HTML, contains the exact same PDF content (Source: Knopf Online, 2004).

have a unique profile: Users with a more flexible learning style most willingly use online help; more structured learners use online help least willingly (Stoddard, 1985). To its credit, the Brio online documentation (Figure 4) eliminates one paragraph of description included in other media (Figures 3 and 5); nonetheless it retains the exact same other structural markers; for example, the same headings and definitions. Other reports have suggested that consumers use print manuals for browsing and gaining an overview and use online help for quick, immediate fixes and for hands-on guidance (Choney, 2002). The online and applet versions of the Brio documentation, however, contain the same content as the PDF version. The online version also uses the same sentences as the PDF version, rather than shorter sentences tailored for e-users. Online help requires those who use it to possess a degree of computer proficiency not required by manuals, and older Web users tend to need more basic online help (Nahm et al., 2004). Yet in the Brio example, the applet (which would be accessed from inside the software it describes and explains) contains the same content as the PDF even though two very different users would be reading these documents.

Unlike hard-copy procedure manuals, online help sites and software can themselves be unfamiliar tools about which users need clarification. As a result, online help generally requires more layers of instruction than print does. Although online help must contain a metalevel of direction indicating why user actions cause specific

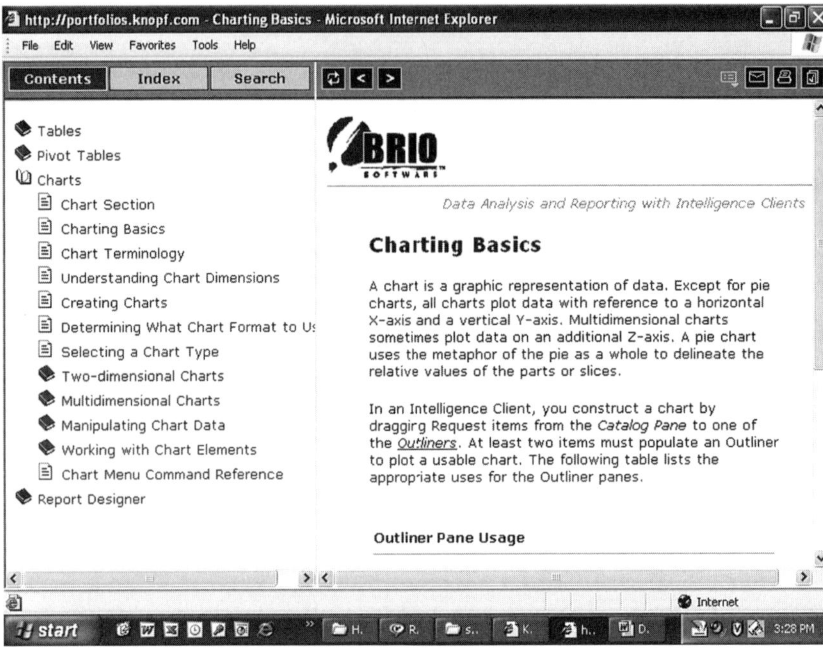

Figure 5. The Brio software applet, containing the same content as the PDF and the Web site (Source: Knopf Online, 2004).

system reactions (Cherry & Jackson, 1989; Purchase & Worrill, 2002; Sutcliffe, 1995), the Brio online documentation (Figure 4) eliminates a paragraph of just such description. Researchers also recommend that online help contain tutorials that do not normally accompany print manuals, and that quick references be offered as teasers to more full explanations (Cherry & Jackson, 1989; Dix, Finlay, Abowd, & Beale, 1993). The Brio online menu (Figure 4) contains no such tutorial, even though much of the information it provides—the key terms, for example—could be better organized and introduced hands-on within a tutorial. Although certain guidelines for print and Web texts are the same, such as write concise passages, use lists, use formatting to aid readability (Gregory, 2004), each medium entails its own best practices. To its credit, Brio's online version (Figure 4) does use chunking (that is, it separates subheadings onto individual Web pages).

While such issues stemming from content management systems and outsourcing are clearly significant concerns for technical communicators, I cannot agree with researchers who point to these difficulties as proofs of a decline in the field (Giammona, 2004). Rather, I argue that technical communicators with a rhetorical orientation are best equipped to step in, oversee, and adjust XML document production, especially given the ever-changing nature of XML protocols (Dumbill, 2001).

In short, we can adjust XML methods in ways that reflect rather than neglect our training in rhetorical theory and human-factors research.

DIVISION OF LABOR AND INTERESTS IN SINGLE-SOURCE DOCUMENTATION

The prior sections of this chapter have demonstrated that XML and single-sourcing projects must not only index content but make it usable. The sections also have suggested how single-sourcing and XML often do not live up to their potentials for tailoring content to specific audiences and genres. These projects all too often reflect a one-size-fits-all approach and treat single-sourcing as an opportunity to reuse rather than to retool content. The following section explores what often causes and aggravates this problem: Namely, unequal knowledge among single-sourcing team members.

The difference between an index generated through XML code and a readable and rhetorically apt index is often the result of a division of labor in single-source documentation. Single-source projects require close working relationships among members of the allied fields that collaborate to produce technical documentation. Stakeholders with differing agendas assume different roles in the process. What is in the best interest of one stakeholder at one stage can often conflict with that of another stakeholder at another stage. Before XML and modern content-management software, writing teams often were expected to write entirely different copy about the same product. For example, it was the manual writer's job to present information about a computer program. Web designers were expected to present data about the same computer program in such a way that Web surfers could access information with ease. And engineers spoke to one another in an entirely different vernacular about the same program.

But XML has made it possible for organizations to create, among other things, user manuals and Web sites from the same master file by imposing modularity on all documents, even those whose content does not lend itself to such a format. Before single-source documentation, writers conceived of their projects as complete units: They wrote chapters of manuals and understood the text they authored in terms of its genre. Single-source writers must conceive of content more fluidly; they must see what they write as having the ability to fit a variety of deliverables in diverse genres, media, and situations. "Modular writing," states Filipp Sapienza (2004), "challenges the assumption to connect ideas to existing or given situations because the writer either does not know where the module will be displayed, does not know in what relation to other content it will be arranged, or must contrive multiple givens" (p. 403).

Modularity entangles the publication process from engineer (or technical professional), to technical communicators and editors, to the intended audience. Because one XML source document often stands as the text for all deliverables, communications and publications departments can no longer assign one book, manual, or deliverable to a writer or editor. Instead, departments divvy up information domains, content segments, or writing tasks (Brown & Walters, 2004; Hackus &

Hedlund, 2001), and each member of the writing team becomes responsible for a subject matter or a particular section to be included in multiple deliverables. Time and turnover can further complicate the objective of maintaining efficient XML schemas for large organizations and projects. In a perfect world, all key personnel would be involved from the beginning to the end of a project as the team made a list of the potential elements for all multiple deliverables and spelled out how these elements were to be related to one another. But, in reality, technical communicators often inherit projects after the fact and must make due with the architecture as conceived by others, whose original goals may no longer be shared by the project team.

Technical communicators may respond to these emerging professional constraints by adopting practices that conflict with their disciplinary understanding of primary and secondary audiences and exigencies. An evolving professional norm for XML coding in the technical communication field calls for composing "media-neutral" content that can be used in multiple different venues and with many different technologies (Rockley, 2001; Ward, 2001). Another norm assumes that communicators will design information that offers content written in a "universally correct" way; that is, so that local and international audiences are approached identically (Jones, 2002). Sapienza defines XML's capacity to provide "different views of the same content from a single source" as *interoperablility* (2004, p. 399). In this view, single-source writers must have multiple audiences, multiple contexts, and multiple media in mind when they write the source content from which will be produced a host of deliverables.

But the difficulty for technical communicators does not lie exclusively in institutional pressures to adopt an arhetorical approach. The document type definition (DTD) defines all elements and attributes and tells a computer or application unfamiliar with an XML document how elements and subelements are related and which attributes are allowed for each element (Applen, 2001, pp. 203–204). Writing and maintaining the DTD is usually not the job of the technical communicators composing the content; that responsibility tends to be reserved for the content analyst or the DTD developer or programmer. Thus, tensions can flare when technical communicators misunderstand the nature of XML and use it for purposes for which it was not designed. Douglas Rudder (2002) describes how certain editors accustomed to print copyediting attempt to use XML tags for formatting (2.1 Potential Mindset Issues section). Often, developers committed to preserving the integrity and simplicity of the DTD will hesitate to accommodate technical communicators' noncritical requests (section 2.1). The friction is reminiscent of the time when team members from different affiliated fields prioritized different objectives for the same deliverables; in some instances, engineers attended more to system issues than to user needs, and technical communicators, vice-versa (Geisler, Rogers, & Haller, 1998, pp. 6–8, 17–18). In short, XML is expected to conform to the syntactic strictures of its DTD, and this exigency can undermine a technical communicator's editorial and creative license (Rudder, 2002, 1 Introduction section).

The tensions between XML indexing and usability, then, stem in part from the divergent interests of technical communicators and their colleagues, between our audience-centric training and the requirements of new media, and between hypothetical work scenarios and the work world itself, in which the composition of single-source teams is often in flux. Certainly, technical communicators and content analysts both have a stake in preserving the integrity of the DTD. It is a mistake, however, to write oversimplified XML schemas in order to maintain that integrity. Tags merely scratch the surface of the context and meaning of information, and it is rhetorically uninformed to act as if format is the main or the only difference between the deliverables of new media.

CORRECTIVES SUGGESTED BY HUMAN COMPUTER INTERACTION (HCI) AND USABILITY TESTING

The trends in single-source composition that I have just described disregard the most crucial tenet of technical communication; namely, communicators' responsibility to produce audience- and genre-sensitive content. In fact, we must apply what we know from rhetorical theory, human-computer interaction research, and usability testing to how we write XML. Although applying rhetorical theory might seem superfluous in light of DTD management, using rhetorical principles effectively in writing code is consistent with the principles of literate programming—a new practice in computer programming. Donald Knuth (1984) introduced literate programming in order to promote the writing of user-friendly computer programs, explaining that, "instead of imagining that our main task is to instruct a *computer* what to do," he meant to "concentrate rather on explaining to *human beings* what we want a computer to do" (p. 1). Both literate programming and single-sourcing emerged in the mid-1990s as a means by which to simplify documentation in computer programming (Stavley, Walsh, & Shipman, 2007, pp. 24–25). Literate programming reflects a belief that programs should be written for humans to read rather than for computers to compile and execute; thus documentation (the instructions describing how and why a program is written) is viewed as just as important as, if not more important than, source code. Literate programming created the means by which to annotate a computer program with detailed descriptions of what each subroutine in the program accomplishes. In essence, it generates a single-source for both a computer program and the manual and explanatory comments about that program. Literate programming has, among its other advantages, freed programmers to write sections of documentation as they write code, often out of chronological order. This chapter (and its promotion of applying rhetorical strategies to coding XML) can be seen as an extension of this method of computer programming, one in which programmers attempt to make artificial languages as user- and human-sensitive as possible.

To bridge the gap between functional XML code and usable and readable XML deliverables, technical communicators can consult research in human-computer interaction and usability testing to learn to create parsing-friendly sentences by creating fine-tuned tags and data categories for the DTD, and, in doing so, help define best practices for attuning XML coders to user needs. Sapienza's research suggests adopting usability measurements as guidelines for XML coding and recommends specifically that XML coders use an information subtypes evaluation (ISTE) when creating subchildren element tags (2004, pp. 404–405). The ISTE example that Sapienza describes focuses on three subtypes of online content that users treat differently: *Conceptual* content, indirectly or tangentially relevant to the task at hand; *coordinative* content, helpful for making transitions from one to another step within that task; and *constructive* content, directly important to it. Were you reading instructions on how to use a particular software, conceptual content would describe the context in which you'd use the software. It would tell word processing software users, for example, the range of writing tasks that the software can perform. Constructive content would offer up specific steps for specific functions. For instance, it would explain to word processing software users that they must click on certain menu options to accomplish certain tasks. Coordinative content, on the other hand, would explain the steps further, by indicating what happens upon the completion of each. For example, it would forewarn word processing software users that a new window with different features will open when they click certain menu options. Sapienza prescribes using XML attributes to denote differences among these content types and applies the corrective to a set of instructions for printing data:

```
<heading>Printing your data</heading>
<instructions subtype="conceptual">You can print your data on paper and on
screen. Printing on screen gives you an impression of what your print on paper
will look like</instruction>
<instruction subtype="coordinative">If you see the PRINT PARAMETERS screen
you are ready to print</instruction>
<instruction subtype="constructivecommand">Move the cursor to Direct
printing.</instruction>
<instruction subtype="constructivecommand">Type Y.</instruction>
<instruction subtype="constructivecommand">Move the cursor to Confirm
action.</instruction>
<instruction subtype="constructivecommand">Type Y.</instruction>
```

(Sapienza, 2004, p. 404)

Here, attributes designate each subtype of data. When creating a document to call up such data, content analysts can use a style sheet or other system to query attributes selectively and (a) format the various levels of data differently; (b) pick and choose which attribute levels to include or abandon, given the audience; and (c) track subtype use (p. 404). The same method also can differentiate among information required of users at different skill levels.

An ISTE is but one usability methodology from human-computer interaction research worth adopting in XML coding. Jakob Nielsen's (2006) heuristic evaluation also is beneficial. In heuristic evaluation, XML writers use specific usability principles to inspect the online help site and devise XML tags (Nielsen, 1994; 2006). One heuristic principle has an especially immediate and obvious application for XML: *Visibility of system status* (VSS), which demands that Web sites provide timely feedback for users (2006). The XML code in this instance requires an additional subclass denoting online help feedback, above and beyond those feedback statements elaborating step-by-step instructions. And XML code that is more media-sensitive can account for even the verbiage of both instructions and meta-instructions.

The iCalibre instructions template reproduced earlier in this chapter would, for example, be improved greatly by the addition of an open-and-close <system-status> tag with which to compose additional metadetails about each step. Consider the following example using an updated iCalibre template that accommodates system status information:

```
<step>
  <cmd >In the File menu, click Preferences.</cmd>
  <info><note>Note: You can use keypad shortcut.</note>Press CTRL and
  F.</info>
  <system-status>A drop-down menu will appear with File command
  options</system-status>
</step>
```

Another heuristic principle—matching between system and real world—would require XML code to account for differences in the experiences of users. For example, since technophobes and technophiles bring different background knowledge to reading and understanding software help manuals, XML coders can devise a scheme of attributes assigning specific theme-rheme combinations for audiences with different degrees of expertise. Earlier in this chapter, the first command step in the iCalibre instructions folded two important pieces of information into the theme:

1. Users must open the file menu in order to
2. select the Preferences option.

This earlier version assumed a level of expertise that older or less experienced users might not have.

Instead, the instructions might append audience-sensitive themes or rhemes as XML attributes:

```
<step>
  <cmd type="beginner">Open the File menu located at the top left of the
  screen.</cmd>
  <info><note>Note: You can use keypad shortcut.</note>Press CTRL and
  F.</info>
  <system-status>A drop-down menu will appear with File command
  options</system-status>
  <cmd type="pro">Select Preferences from the File menu window.</cmd>
</step>
```

The revised version uses XML attributes to distinguish among audiences of differing proficiencies and assumes that an expert does not need to be instructed in as many steps as a beginner does. The DTD and a style sheet, then, can create a beginner's manual by importing instructions indicated by the "beginner" attributes, and an advanced manual from importing "advanced" attributes.

In addition, XML writers can be helped in producing more audience-tailored code by using claims analysis, a strategy developed by John Carroll and Mary Beth Rosson (1992). This activity involves a team of XML writers brainstorming scenarios of use or psychological effects of the site's features on users. Then, in scenario-based design, solutions are composed for each hypothetical scenario. All features of a scenario-sensitive Web site are rationalized. Claims analysis works much like an organized brainstorm that produces comparatively fine-grained XML tags and attributes (like the ones described earlier in this section). As a result, the Web site embodies claims or assumptions that designers have about the strengths and weaknesses of the site.

CONCLUSION

Although the XML methods suggested in this chapter might not solve industry problems of competing interests, inherited projects, or overcomplicated DTDs, industry pressures do not render these suggestions implausible or impossible. Clearly, it is important to have XML-proficient technical communicators making decisions and collaborating in key positions at key stages in a single-sourcing project—from beginning to end, from management to staff.

To date, the technical communication field has provided writing and editing consultation and services to a variety of fields. Our services have included writing manuals and user documentation accompanying programs, products, and services. These are obvious entry points where rhetorical skill is vital, and content writing (generating the text for multifold genres and media) will continue to require our expertise. Emerging technologies such as XML, however, give rise to new situations. Contemporary forms of content management and information architecture, or the complex of decisions regarding how to catalog, organize, and disseminate content, are symptomatic of a complex communicative situation that is more demanding than ever of our rhetorical acumen.

Single-source documents reflect the challenging rhetorical situations for which they are written: They are consumed by multiple, disparate audiences in various dissimilar contexts constrained by different levels of reading experience and environments of use. In contrast, the prevalent approach to single-sourcing in industry reflects a structuralist understanding of audience: A self-similar group of users or readers with shared expectations for the documentation, and a group who will use the documentation in similar ways. Professional applications of XML in single-sourcing projects operationalize this structuralist understanding by prescribing and preparing essentially the same content for every incarnation excerpted from the single source of content.

As a corrective, we must employ the poststructural rhetorical and usability resources at our disposal, and our contribution to single-sourcing projects must be to dissect and to manage content in ways that acknowledge the polyphonic nature of the single-source document. As Sapienza (2007) describes, because a single-source document must comprehensively anticipate the incarnations and contexts in which its bits and pieces will be used, the writing and preparation process is unlike that for most other preexisting genres of documentation (e.g., instructions, Web sites) (p. 87). Thus, we must work differently from the ways that we work in the standard genres, which reflect the specific constraints evolving out of readers' expectations and patterns of use over time. In single-sourcing projects, the composing process calls for more granularity (or finding the smallest units of content and usable information) and subtextuality (or writing with full consciousness of other, related texts pertinent to the text) (p. 92). Enacting this process will require more than simply writing multiple iterations of text; it will require prescribing, creating, and delimiting the various elements, attributes and subsets of content necessary for addressing and forecasting the needs of several awaiting audiences. In this light, rhetorical interventions in single-sourcing can highlight the indispensable abilities of technical communicators.

REFERENCES

Applen, J. D. (2001). Extensible markup languages and traditional abstracting and indexing strategies. *IEEE Transactions on Professional Communication, 44*(3), 292–296.

Bazerman, C. (1994). Systems of genre and the enactment of social intentions. In A. Freedman & P. Medway (Eds.), *Genre and the new rhetoric* (pp. 79–99). London: Taylor & Francis.

Brown, L., & Walters, C. (2004). Shacking up with single sourcing: Making the relationship work. *Currents 2004: Proceedings from the Society for Technical Communication.* Atlanta: STC Atlanta Chapter. Retrieved October 1, 2006, from http://www.stcatlanta.org/currents04/proceedings/brown_walters.pdf

Carroll, J. M., & Rosson, M. B. (1992). Getting around the task-artifact cycle: How to make claims and design by scenario. *ACM Transactions on Information Systems, 10*(2), 181–212.

Cherry, J. M., & Jackson, S. R. (1989). Online help: Effects of content and writing style on user performance and attitudes. *IEEE Transactions on Professional Communication, 32*(4), 294–299.

Craft, J. (2004). *XML as a computational and rhetorical technology*. Computer Writing and Research Lab, University of Texas at Austin [White paper]. Retrieved October 1, 2006, from http://www.cwrl.utexas.edu/files/040512-1.pdf

Choney, S. (2002, September 16). Computer users, makers not on the same page. *San Diego Union-Tribune*. Retrieved at http://findarticles.com/p/articles/mi_hb5262/is_200210/ai_n20384557

Dix, A., Finlay, J., Abowd, G., & Beale, R. (1993). *Human computer interaction*. London: Prentice Hall.

Dumbill, E. (2001, May 31). The state of XML: Why individuals matter. *XML.com*. Retrieved November 30, 2005, from http://www.xml.com/pub/a/2001/05/30/stateofxml.html?page=2

Geisler, C., Rogers, E. H., & Haller, C. (1998). Disciplining discourse: Discourse socialization in the affiliated professions of software engineering design. *Journal of Written Communication, 15*, 3–24.

Giammona, B. (2004). The future of technical communication: How innovation, technology, information management, and other forces are shaping the future of the profession. *Technical Communication, 51,* 349–367.

Gregory, J. (2004). Writing for the Web versus writing for print: Are they really so different? *Technical Communication, 51*, 276–285.

Hackus, J., & Hedlund, T. (2001). *Making a business case for single sourcing.* The Center for Information-Development Management. Retrieved October 1, 2006, from http://www.comtech-serv.com/pdfs/Business_Case_for_Single_Sourcing.pdf

Halliday, M. A. K., & Matthiessen, C. M. I. M. (1994). *An introduction to functional grammar.* London: A. Hodder Arnold.

Jones, A. (2002). *Localization of single-source content.* Society for Technical Communication 2002 Annual Conference Proceedings. Arlington, VA: Society for Technical Communication. Retrieved October 1, 2006, from http://www.stc.org/confproceed/2002/PDFs/STC49-00043.pdf

Knopf, D. (2004). *Knopf online portfolio: Single source solutions.* Retrieved February 28, 2006, from http://www.knopf.com/portfolio/single_source.html

Knuth, D. E. (1984). Literate programming. *The Computer Journal, 27*(2), 97–111.

Mirel, B. (1991). Critical review of experimental research on the usability of hard copy documentation. *IEEE Transactions on Professional Communication, 34*(2), 109–122.

Nahm, E. S., Preece, J., Resnick, B., & Mills, M. E. (2004). Usability of health Web sites for older adults: A preliminary study. *CIN: Computers, Informatics, Nursing, 22*(6), 326–334.

Nielsen, J. (1994). *Enhancing the explanatory power of usability heuristics.* CHI '94: Celebrating Interdependence, Human Factors in Computing Systems (pp. 152–158). Boston, MA: Association for Computer Machinery.

Nielsen, J. (2006). *How to conduct a heuristic evaluation.* Retrieved March 15, 2006, from http://www.useit.com/papers/heuristic/heuristic_evaluation.html

Purchase, H. C., & Worrill J. (2002). An empirical study of on-line help design: Features and principles. *International Journal of Human-Computer Studies, 56*, 539–567.

Rappaport, L. (2002, December). XML takes on a documentation nightmare. *Transform Magazine*. Retrieved October 1, 2006, from http://www.transformmag.com/db_area/archs/2002/12/tfm0212ic.shtml

Rockley, A. (2001). The impact of single sourcing and technology. *Technical Communication, 48*(2), 189–193.

Rockley, A., & Manning, S. (2000). *XML and single sourcing.* Society for Technical Communication 2000 Annual Conference Proceedings. Arlington, VA. Retrieved October 1, 2006, from http://www.stc.org/confproceed/2000/ PDFs/00120.PDF

Rudder, D. (2002, May). Technical writing and XML: Reconciling editorial license with structured markup. *XML Files: The XML Magazine, 37*. Retrieved October 1, 2006, from http://www.idealliance.org/xmlfiles/issue37/techwriting.asp

Sapienza, F. (2004). Usability, structured content, and single sourcing with XML. *Technical Communication, 51*(3), 399–408.

Sapienza, F. (2007). A rhetorical approach to single-sourcing via intertextuality. *Technical Communication Quarterly, 16*(1), 83–101.

Scott, L. M. (1994). The bridge from text to mind: Adapting reader-response theory to consumer research. *The Journal of Consumer Research, 21*(3), 461–480.

Stavely, A., Walsh, L., & Shipman, J. (2007). Lightweight literate programming: A documentation practice. *Technical Communication, 55*(1), 23–37.

Stoddard, M. L. (1985). *Learning styles and embedded training: A case study* (report no. IR 012 276). Los Alamos, NM: Los Alamos National Laboratory, U.S. Department of Energy. Washington, DC. (ERIC Document Reproduction Service No. ED 274 305)

Sutcliffe, A. (1995). *Human computer interface design* (2nd ed.). Houndmills, UK: Macmillan.

Vasudev, P., & Patel, R. (2006). *Writing for DITA. iCalibre.* Retrieved February 10, 2006, from http://www.icalibre.com/mambo/index.php?option=com_content&task=view&id=38&Itemid=47

Ward, N. (2001). *Are you ready to repurpose? The challenge is to separate content from form.* Retrieved October 1, 2006, from http://www.ondemandpublishing.com/QPP_Repurposing.html

Yates, J., & Orlikowski, W. (2002). Genre systems: Structuring interaction through communicative norms. *Journal of Business Communication, 39*(1), 13–35.

Contributors

FAIZA DERBEL is Assistant Professor and Chair of English and Linguistics at the Faculty of Letters, Arts, and Humanities, University of Manouba, Tunisia. She teaches graduate and undergraduate courses in various areas of applied linguistics and is engaged in research on technology and language pedagogy, teacher cognition, and interethnic/intercultural communication. In fall 2008, she was a Fulbright Research Fellow at Oregon State University, where she engaged in scholarship leading toward the completing of a book on digital pedagogy, tentatively titled *Taking the Technology Route: Meeting Educational Challenges in a Globalizing World.* She is currently leading a group of four designers of an online Business English national curriculum for the Virtual University of Tunisia.

HUILING DING is Assistant Professor of Professional Communication at Clemson University. Interested in professional communication, disciplinary writing, medical rhetoric, and comparative rhetoric, she has published in journals such as *Written Communication, Business Communication Quarterly, Rhetoric Review,* and *English for Specific Purposes*.

JASON FARMAN has published in the areas of embodiment and space in digital environments, with recent articles on the topics of digital performance, videogame interfaces, and the computer virus. This research earned him the Chancellor's Fellowship for Dissertation Year Research at UCLA and a research grant through the University of California Santa Barbara's Transliteracies Project. He currently is working on a book manuscript titled *Mobile Interface Theory,* to be published by Routledge in 2011. In fall 2010 he will join the American Studies faculty at the University of Maryland, College Park.

WILLIAM FAUX is Associate Professor of Speech Communication at Valdosta State University. He teaches in the areas of organizational communication and intercultural communication. Dr. Faux has studied the cultural identity of anthracite coal miners of northeastern Pennsylvania and is conducting research on the impact of accent in interpersonal communication.

JOHN B. KILLORAN is Assistant Professor in the English Department at Long Island University, Brooklyn campus. He researches Web communication and has published in journals including the *Journal of Business and Technical Communication*,

IEEE Transactions on Professional Communication, the *Journal of Technical Writing and Communication*, and *Computers and Composition*.

HEEMAN KIM is Assistant Professor of Communication at Kennesaw State University. He teaches communication research methods and contemporary issues in mass media. His research includes the cultural and social uses of communication technology and international communication, with an emphasis on Asian and U.S. contexts.

RUDY MCDANIEL is Assistant Professor of Digital Media at the University of Central Florida. His research interests include interactive technical communication, narrative theory, video game technologies, knowledge management frameworks, and extensible markup language (XML). He is co-author of *The Rhetorical Nature of XML: Constructing Knowledge in Networked Environments* (Routledge, 2009). As a consultant, he has designed Web-based applications for clients such as the IEEE Society and the Library of Congress and has written interactive video-game scripts for the Federation of American Scientists. He is currently director of the Partnership for Research on Storytelling Environments (PROSE) lab at UCF.

LAURA MCGRATH is Associate Professor of English at Kennesaw State University in the Atlanta, Georgia, area. She teaches a variety of rhetoric/composition and professional writing courses. As a technology specialist, Laura's scholarship and service activities focus on online learning, new media, writing in digital environments, and faculty development and support. She currently is editing a collection titled *Collaborative Approaches to the Digital in English Studies,* under consideration at the Utah State University Press imprint Computers and Composition Digital Press.

AIMEE KENDALL ROUNDTREE is Assistant Professor in the Professional and Technical Writing Program at the University of Houston-Downtown, where she teaches courses in hypermedia and technology, Web publishing and accessibility, visual design, medical and science writing, and technical communication. Her research interests examine scientific discursive practices, rhetorical implications of new media, visual design, and information technology. She has published essays on the rhetoric of scientific invention, pedagogy in MOOs and virtual space, and mimesis in computer simulations. She is currently working on a book about the rhetorical nature of computer simulations and a research project examining cancer survivorship blogs.

EILEEN E. SCHELL is Associate Professor of Writing and Rhetoric and Chair and Director of the Writing Program at Syracuse University. She is the author of *Gypsy Academics and Mother-teachers: Gender, Contingent Labor, and Writing Instruction* (Heinemann, 1998), the co-editor with Patti Stock of *Moving a Mountain: Transforming the Role of Contingent Faculty in Composition Studies and Higher Education* (NCTE, 2001), and co-author with Kim Donehower and Charlotte Hogg of *Rural Literacies* (Southern Illinois University Press, 2007). Her co-edited volume with Kelly Rawson *Rhetorica in Motion: Feminist Rhetorical Methods and Methodologies* was published by University of Pittsburgh Press in 2010. Her work has appeared in *College English*, *College Composition and Communication*, *JAC*, and *WPA: The Journal of Writing Program Administration.*

SHERRY STEWARD is a simulation and training professional with the Department of Defense. She teaches a variety of courses in rhetoric/composition and technical communication. Her research interests include supportability engineering, technical communication, narrative theory, video-game technologies, and interactive electronic technical data.

LEONARD WITT holds the Robert D. Fowler Distinguished Chair in Communication at Kennesaw State University, located outside of Atlanta. He is founder and executive director of the Center for Sustainable Journalism, funded by a $1.5 million gift from the Harnisch Foundation. The Center nurtures and disseminates innovative ideas, produces research, incubates new projects, and launches applications, all aimed at producing financially sustainable, high-quality, ethically sound journalism. He is founding president of the Public Journalism Network (PJNet.org) and past chair of the Civic and Citizen Journalism Interest Group in the Association for Education in Journalism and Mass Communication (AEJMC). He was a magazine and features editor for more than 25 years, including 6 years as *Sunday Magazine* editor with the *Minneapolis Star Tribune* and 6 years as editor of *Minnesota Monthly* magazine. Before entering academia in 2002, he was the executive director of the Minnesota Public Radio Civic Journalism Initiative.

Index

Abbott, Christine, 190
Academy, digital technology compromising foundations of the, 8, 198
See also Broadband divide and technical communication pedagogy; Résumés on the Web, self-published; *individual subject headings*
Adams, Jonathan, 10
Adobe products, 181
Advocacy. *See Meatrix, The;* News delivery in an era of digitally mediated democracy; Textual restructuring as encouraged by digital forms
Africa and mobile phones, 4-5
Agriculture, United States Department of (USDA), 32
See also Meatrix, The
Alienation Effect, 83, 89
Allen, Nancy, 180, 181
AltaVista, 100, 103
Altbach, Philip, 8
American Society of Magazine Editors, 10
AOL, 164
Appadurai, Arjun, 139-140
Apple Computers, 42
Association of Moving Image Archivists, 63
Asynchronous communication, 162, 170
Auslander, Philip, 79
Authorial norms/authority and textual restructuring as encouraged by digital forms, 46-52
Bakersfield Californian, 65, 70
BaristanetNJ, 60
Barwell, Graham, 162
Bayosphere, 60
Beloff, Zoe, 85, 88, 89
Benjamin, Walter, 208
Benkler, Yochai, 58, 65, 67, 68, 70
Benninghoff, Steven, 180, 181
Beongae events, 158
Binary oppositions, 2, 3
Blackboard, 9
Bollier, David, 26
Bolter, Jay D., 42, 89-90
Bolton, Ruth, 10
Bosley, Deborah S., 139
Bourdieu, Pierre, 11-12, 95, 96-99, 107, 112
Bousquet, Marc, 23
Bowles, Kate, 162
Brandt, Deborah, 24
Brecht, Bertolt, 83, 89
Brio, 221-225
British Broadcasting Corporation (BBC), 63
Broadband divide and technical communication pedagogy
academic perspectives, 197-198, 200-202
creative skill gaps, 204
curriculum (proposed) for technical communication, 206-211
deficiencies in professional entry-level skills for technical communicators, 209-210

[Broadband divide and technical communication pedagogy]
education programs and business, fissure between, 196-199
gap between academic and industrial thought, 196-199
group dynamics, 203-204
holistic view, 201
industrial perspectives, 202-206
Interactive Electronic Technical Manual, 202-203
interactivity, computational, 210
internships, collaborative, 210-211
management and leadership courses, 211
multidisciplinary education, 210
multimedia-enabled processes, 208
multimodal approach, 198-200, 207-208
overview, 195-196
positivist knowledge paradigm, 196
reconciling academic and industrial objectives, 206-209
summary/conclusions, 211
writing, academic focus on traditional, 199
Bromme, Rainer, 44
Brooke, Robert, 20
Brown, Lori, 217
Brown, William, 160
Buckland, Michael, 47-49
Bureau of Labor Statistics, 100
Burnett, Rebecca, 139
Bush, Vannevar, 47
Byram, Michael, 135

Callens, Johan, 83
Carey, James, 64
Caribbean, 5
Carliner, Saul, 190
Carroll, John, 230
Case, Sue-Ellen, 86
"Case for 'Technical Communicator'" (Martin & O'Sullivan), 181
"Cathedral and the Bazaar, The" (Raymond), 57, 73
CD-ROM, interactive. *See* Performance in the digital age
Cell phones, 4-5
Central Intelligence Agency (CIA), 59
Cheon, Hongsik, 161
China, 7, 137
Cho, Chang-Haon, 161
Choo, Chun W., 120
Citizen media. *See* News delivery in an era of digitally mediated democracy
Citizens' Environmental Coalition, 31
Civic mapping, 62
Coase's Penguin (Benkler), 58
Collaboration. *See* Partnerships between businesses/organizations/communities
Collectivism *vs.* individualism, 159-161
See also Friends offline, cultural/individual differences and meeting online
Commons-based peer production. *See* News delivery in an era of digitally mediated democracy
Communication technology. *See* Friends offline, cultural/individual differences and meeting online
Communication World, 182
Communities, building. *See* Friends offline, cultural/individual differences and meeting online; South-North collaboration between professional writing students
Composition and Sustainability (Owens), 20
Computer-mediated communication (CMC), 158-159
See also Friends offline, cultural/individual differences and meeting online
Concentrated animal feeding operations (CAFOs), 25, 28, 30, 32
Conceptual content, 228
Conference on College Composition and Communication (CCCC), 125
Constructive content, 228
Content and information subtypes evaluation, 228
Content management systems, 213
Cook, Kelli C., 9, 185
Cooper, Athena, 104
Coordinative content, 228
Copyright issues and citizen journalism, 63
Creative Commons, 63

Creativity and the broadband divide, 204
Cultural capital, 96-97
  See also Résumés on the Web, self-published
Cultural/individual differences. *See* Friends offline, cultural/individual differences and meeting online; South-North collaboration between professional writing students
Cultural knowledge, 120
Cultural variability, Hofstede's theory of, 172
Cushman, Ellen, 188
Cyberactivism. *See* Advocacy

Damoim, 164
Daum Café, 164
Dautermann, Jennie, 8
Davis, Marjorie, 205
Dayton, David, 182
Deane, James, 3
DeFleur, Margaret, 10
Delany, Paul, 41
Democracy, digitally mediated. *See* News delivery in an era of digitally mediated democracy
Derbel, Faiza, 12, 137, 138, 235
Designers/users binary, 6-7
DeVoss, Danielle N., 188
Dias, Patrick, 36
Digital divide, the, 2-6
  See also Divergence/complexity, digital
Digital Future Initiative Panel, 63
Digital Nation (Wilhelm), 22-23
Ding, Huiling, 12, 235
Discourse model, Kress and Van Leeuwen's, 199
Discovery of Grounded Theory, The (Glaser & Strauss), 121
Dispositio (arrangement of that which has parts). *See* Textual restructuring as encouraged by digital forms
Disruptive technology, 57, 58
  See also News delivery in an era of digitally mediated democracy; Textual restructuring as encouraged by digital forms
Distribution systems and citizen journalism, 72
Divergence/complexity, digital
  academy, digital technology compromising foundations of the, 8
  binary oppositions, 2, 3
  credibility of all digital-media businesses, 10
  cyberactivism, 21-23
  divide, the digital, 2-6
  English studies, entrenchment of digital technology in field of, 1
  governance, public/corporate, 7-8
  marketing and public relations, 10-11
  overview, 1-2
  pedagogy, professional communication, 9-10
  South-North collaboration between professional writing students, 145-150
  users/designers, 6-7
  See also Broadband divide and technical communication pedagogy; *individual subject headings*
Divide, the digital, 2-6
  See also Divergence/complexity, digital
Dobrin, Sidney, 187
Doctor Faustus Lights the Lights (Stein). *See* Performance in the digital age
Document type definition (DTD), 226, 227, 230
Dreamweaver, 139, 181
Duin, Ann H., 6, 8, 23, 191

Eble, Michelle, 23
Economic differences and the digital divide, 4
Ecopoetic theory, 20
Edmondson, Jacqueline, 29
Education and industry, 8, 198
  See also Broadband divide and technical communication pedagogy; Résumés on the Web, self-published; *individual subject headings*
EDUCAUSE Learning Initiative, 1
Edwards, J. D., 214
Eiler, Mary A., 41
Elements of Style, 38
Emanuel Goldberg and His Knowledge Machine (Buckland), 48
Embodied cultural capital, 99

English studies, entrenchment of digital technology in field of, 1, 134
Entica, 164
Ethnocentrism, 157-158
European Union, 137
"Evaluating the Intellectual Work of Writing Administrators," 122
Extensible markup language (XML). *See* XML documentation, rhetorical interventions for

Faber, Brenton, 37
Factory Farms in the State of New York, 31
Fallows, Deborah, 96
Farkas, David, 45
Farman, Jason, 11, 49, 235
Farming, factory. *See Meatrix, The*
Fast Food Nation (Schlosser), 25, 30-31
Faudree, Roy, 84
Faux, William, 12, 21, 72, 235
Fence and Windows (Klein), 21, 24
Ferguson, Douglas, 164
Findlay, Jim, 83
First Monday, 57
Flaherty, Lisa, 161
Flash films. *See Meatrix, The*
Food and Drug Administration (FDA), 32
Food consumption. *See Meatrix, The*
Fourth Estate. *See* News delivery in an era of digitally mediated democracy
FrameMaker, 181, 220
Freedman, Aviva, 36
Freedman, Samuel, 64
Free Range Studios, 25-26
Free Software Foundation, 62
Friends offline, cultural/individual differences and meeting online
  access, Internet, 162
  asynchronous/synchronous communication, 162, 170
  Beongae events, 158
  collectivism *vs.* individualism, 159-161
  computer-mediated communication, 158-159
  ethnocentrism, 157-158
  methods, study, 165-166
  overview, 157
[Friends offline, cultural/individual differences and meeting online]
  participation in social/political networks, increased, 158
  portal services, 164
  research on cultural uses of the Internet, 161-162
  results, study, 166-171
  self, independent/interdependent, 160-164, 170-172
  summary/conclusions, 171-172
From Memex to Hypertext (Nyce & Kahn), 47
Fulbright programs, 8
Furuta, Richard, 42

Gaillet, Lynee L., 23
Gaming community, computer, 6-7
Gender and the Internet, 95, 101
Geocities, 103
"Gertrude Stein in QuickTime" (Farman), 49
  See also Performance in the digital age
Gestalt-shift genres, 1
Gillmor, Dan, 60, 61
Global Contexts: Case Studies in International Technical Communication (Bosley), 138-139
Globalization. *See* Friends offline, cultural/individual differences and meeting online; South-North collaboration between professional writing students
Global Resource Action Center (GRACE), 25, 28, 30, 31
Goldberg, Emanuel, 48
Gordon, Jay, 46
Governance, public/corporate, 7-8
Grabill, Jeff, 188
Graham, S. Scott, 1
Granularity and citizen journalism, heterogeneous, 67-68
Grice, Roger, 205
Grocery Store Wars, 25
Gross domestic product (GDP), 137
Group dynamics and the broadband divide, 203-204
Grove, Laurel, 200

Grusin, Richard, 89-90
Gurak, Laura, 8, 191

Hackos, Joann, 20
Halasz, Frank, 48
Halliday, M. A. K., 219
Hanappe, Peter, 62
Hands On Network, 9
Hanmir, 164
Hart-Davidson, William, 183
Hawisher, Gail, 50
Hawk, Byron, 7
Haye, Bethany, 85
Hayes, Elizabeth, 43
Haythornthwaite, Caroline, 170
Heald, Gary, 10
Herring, Susan, 101
Heterogeneous granularity and citizen journalism, 67-68
Hofstede, Geert, 159, 172
Holistic view of product design, 201
Homophily, 4
Hongladarom, Soraj, 163
Horizon Report, 192
Horrigan, John, 96
House/Lights. See Performance in the digital age
H20town, 60
Human-computer interaction (HCI) and usability testing, 227-230
Hunsinger, R. Peter, 139
HyperCard, 41, 42-43
Hypertext
  exploratory *vs.* constructive, 9
  HyperTEXT '87, 48
  Intertextuality and the exploration of hyperlinks, 46
  markup language (HTML), 215-216
  Trellis hypertext model, 42
 See also Textual restructuring as encouraged by digital forms

iCalibre, 218-221, 229
"Impact of the Internet and Digital Technologies on Teaching and Research in Technical Communication" (Gurak & Duin), 191
Individualism *vs.* collectivism, 159-161
 See also Friends offline, cultural/individual differences and meeting online
Industry and the academy, 8, 198
 See also Broadband divide and technical communication pedagogy; Résumés on the Web, self-published; *individual subject headings*
Informatics of domination, 23
Information and communication technology (ICT), 4-5
 See also Technical and professional communication educators; *individual subject headings*
Information subtypes evaluation (ISTE), 227-228
Institutionalized cultural capital, 99
"Integrating the Web into Education for Technical Communication Majors" (Carliner), 190
Interactive art/performance. *See* Performance in the digital age
Interactive Electronic Technical Manual (IETM), 202-203
Intercultural communicative competence. *See* South-North collaboration between professional writing students
International Association of Business Communicators, 183
Internet, the, 5, 6
 See also Friends offline, cultural/individual differences and meeting online; Hypertext; Résumés on the Web, self-published; *individual subject headings*
Interoperability and XML documentation, 226
Intertextuality and the exploration of hyperlinks, 46
Investing, socially responsible, 8
Irigaray, Luce, 86

Jackson, Timothy A., 134
Jenkins, Henry, 50
Johnson, Nicholas, 8
Johnson-Eilola, Johndan, 43
Journalism. *See* News delivery in an era of digitally mediated democracy

Kahn, Paul, 47
Kairos, 35, 39, 43, 52
See also Textual restructuring as encouraged by digital forms
Kalmbach, Jim, 43
Kastman-Breuch, Lee-Ann, 181, 182
Keller, Christopher, 187
Kelly, Kevin, 61
Kelly, Tim, 4-5
Kent, Thomas, 52
Killingsworth, M. Jimmie, 20
Killoran, John B., 11, 51, 235-236
Kim, Heeman, 12, 21, 72, 236
Kim, Min S., 160
Kimball, Miles, 112
Kinder, Marsha, 88
Kitayama, Shinobu, 160
Klein, Naomi, 21, 24
Knopf Online, 221
Knowledge, informal *vs.* formal, 119-120
Knuth, Donald, 227
Kolb, David, 51
Kope, Maryann, 44
See also Friends offline, cultural/individual differences and meeting online
Kouper, Inna, 101
Kress, Gunther, 1, 198, 199, 207-208

Labor Department, U.S. Department of, 100
Lamberti, Adrienne P., 11, 201
Landow, George, 41
Lasn, Kalle, 26
Latin America, 5
Leadership and citizen journalism, 70
LeCompte, Elizabeth, 82
Leeuwen, Theo V., 198, 199, 207-208
Lenhart, A., 101
Lenhart, Amanda, 96
Lessig, Lawrence, 63
Linux, 63, 66, 71
Literate programming, 227
Liveness (Auslander), 79
Localization, professional writing efforts at, 20
Longo, Bernadette, 134
Los Angles Times, 58, 60, 68
Lovelady, Steve, 59
Lycos, 164

Madden, Mary, 101
Marketing and digital divergence/complexity, 10-11
Markus, Hazel, 160
Marlowe, Christopher, 81
Martin, Maurice, 181
Martin, Robert P., 20
Matthiessen, Christian, 219
Mawra, Joseph, 82
Maxmin, James, 10
McDaniel, Rudy, 12, 90, 182, 236
McGrath, Laura, 12, 236
Meatrix, The
agro-industrial complex, 20
call to action, 28-29
classroom, using the film in the, 30-32
corporations, nameless/faceless, 19-20
critical and agricultural literacy, 23-24
cyberactivism and digital divergence, 21-23
educational/organizing tool, effective, 28
oppositional consciousness and alternative solidarities, 19, 23
overview of the movie, 27
place-conscious education, 20
possibilities/limitations of Flash films/nonprofit Web sites, 29-30
summary/conclusions, 32-33
viral Flash activism, 25-26
virus shell of pop culture, 27-28
Media, the. *See* News delivery in an era of digitally mediated democracy
Medway, Peter, 36
Meyer, Phil, 59
Microsoft Office, 181
Middle East, 5
Millennium Development Goals, 4
Miller, Carolyn, 1, 196
Miller, Judith, 59
Miller, Thomas, 45
Minnesota (MN) Public Radio's Public Insight Journalism project, 68
Mirel, Barbara, 8
Mobile phones, 4-5
Modular projects and citizen journalism, 67

Modular writing and XML documentation, 225-226
Moulthrop, Stuart, 42, 47
MSN, 164
Multiliteracies for a Digital Age (Selber), 180
Multimodal communication, 198-200, 207-208
Multitasking, 50
Multivocality, 51
Murray, Janet, 87

National Council of Teachers of English, 1, 24
National Internet Development Agency of Korea (NIDAK), 164
National Rifle Association (NRA), 26
Nation Online, A: Entering the Broadband Age, 195
Natural Resources, U.S. Department of, 32
Naver, 164
Nelson, Ted, 48
Netian, 164
"Networking the Nile: Technology and Professional Development in Egypt" (Warschauer), 2-3
New Media Consortium, 1, 192
News delivery in an era of digitally mediated democracy
  attitude shift for journalists, 61
  citizen contribution extent and effort, 66-68
  citizen involvement that fits your newsroom, 65-66
  commons-based peer production, 58
  disruptive technology, 57, 58
  distribution systems, 72
  ecosystem, draw an, 62
  erosion of mass news-media dominance, 59
  experiments in open-source journalism/citizen outlets, 60
  failure, experiment and increase tolerance for, 71-72
  invited/excluded, deciding who gets, 68
  leadership, 70-71
  newsroom offices or not, 71
[News delivery in an era of digitally mediated democracy]
  nonconsumers rather than core audience, innovations aimed at, 62
  objectivity and other ethics of professional journalism, 64-65
  open-source philosophy, 57-58
  overview, 57
  ownership model, 70
  philosophy and mission check, 62-64
  shift in news media's attitudes toward the digital, 59
  summary/conclusions, 72-73
  timing, rethinking issues of, 69
  transparency issues, 69-70
Newspaper Next: Blueprint for Transformation, 62
New York Times, 59
Nielson, Jakob, 48, 228
Northwest Voice, 60, 65, 70
Nostalgic Angels (Johnson-Eilola), 43
Noveck, Beth, 62, 71
Nyce, James, 47

Objectified cultural capital, 99
O'Dowd, Robert, 144
OhmyNews, 60, 65, 68
Olga's House of Shame, 82
On Rhetoric (Aristotle), 46
Open Source Initiative (2007), 57
  See also News delivery in an era of digitally mediated democracy
"Order and Coherence in a New Medium" (Slatin), 43
Organic Fair Trade Association, 25
O'Sullivan, Richard, 181
Otlowski, Marcus, 42
Overholser, Geneva, 59
Owens, Derek, 20
Ownership model and citizen journalism, 70

Pacey, Arnold, 159
Palvetzian, Talene, 23
Papper, Carole C., 183
Paran, 164
Pare, Anthony, 36
Participatory culture, 50

Partnerships between businesses/organizations/communities, 9-10, 191-192
  See also South-North collaboration between professional writing students; Writing Program Administrators electronic mailing list; *individual subject headings*
PC Bang, 164, 170
Peace Corps, 8
Pearce, Kevin, 161
Pedagogy, professional communication, 9-10
  See also Broadband divide and technical communication pedagogy; Technical and professional communication educators; XML documentation, rhetorical interventions for; *individual subject headings*
Penka, Jeffrey, 10
Performance in the digital age
  blurring lines between live and digitized events, 80-81
  distance created between actor and character, 82-83
  Doctor Faustus Lights the Lights, 81
  event and document, relation between, 79-80
  House/Lights, 80, 82
  interactive performance on the digital stage, 86-89
  overview, 79
  RCA Records' "His Master's Voice" dog, 90-91
  resisting immersive interactivity, 89-91
  spectator-user, involving the, 80, 88
  summary/conclusions, 92-93
  technology as simulator/representor of the "real," 83-86
Perse, Elizabeth, 164
Pew Commission on Industrial Farm Animal Production, 20
Phelan, Peggy, 92
Phelps, Louise, 120
Place-conscious education, 20
Plame, Valerie, 59
Politics of Information, The (Bousquet), 23
Portland Resolution, 119
Poulakos, Takis, 182
Printcentric critiques of digital technologies, 37
  See also Textual restructuring as encouraged by digital forms; Writing Program Administrators electronic mailing list
Profit, symbolic, 98
  See also Résumés on the Web, self-published
Project Xanadu, 47, 48
Public and Professional Writing (Surma), 51
Public Broadcasting Service (PBS), 63
Public Eye, 69
Public participation in debates on use of digital technology, 8, 22
Public relations and digital divergence/complexity, 10-11
Publishing personal sites on the Web, 95-96
  See also Résumés on the Web, self-published

QuestionPoint, 10
QuickTime. *See* Performance in the digital age
Qwest, 9

Rainey, Kenneth, 182
Raisinghani, Mahesh, 20
Raymond, Eric, 57, 62, 63, 69, 70-71, 73
RCA Records' "His Master's Voice" dog, 90-91
Relationships. *See* Friends offline, cultural/individual differences and meeting online
Remediation: Understanding New Media (Bolter & Grusin), 89-90
Resolution on Media Literacy, 1
Resources in Technical Communication (Self), 186
Résumés on the Web, self-published
  age distribution, 102, 114
  Bourdieu on symbolic profit and cultural capital, 97-99
  challenges in finding documents worthy of publishing on the Web, 110
  computer skills, 103

[Résumés on the Web, self-published]
critical perspective on cultural capital, students', 110-111
cultural capital, 96-99
displayable on the Web, cultural capital that is, 103-105
educational distribution, 102, 103
efficacy of, 103, 109, 114
gender differences, 101
genre-rich Web sites, being directed to, 106-107
limitations in acquiring digitizable cultural capital, 110
methods, study, 99-101
narrow range of occupational clusters, 96
new cultural capital, developing, 110-112
occupational distribution, 96, 102, 103, 109, 113
overview, 95
pedagogical implications, 109-112
quantitative indicators of digitized cultural capital, 107-109, 114
skills, differences in Web authoring, 96
Rheingold, Howard, 21
Rhetorical Education in America, (Socrates) 45
Rhetorical Refusals (Schilb), 36
Rice, Rich, 183
Richards, Ann E., 12, 138
Rogers, Everett M., 4
Rose, Shirley, 119
Rosson, Mary B., 230
Roundtree, Aimee K., 12-13, 182, 198, 236
Royal, Cindy, 95
Rubin, Rebecca, 161
Rudder, Douglas, 226
Rule-based knowledge, 120
Rushkoff, Douglas, 26

Sacramento Bee, 59
Saltz, David, 87
Salvo, Michael, 49, 51-52, 186
Sapienza, Filipp, 225-226, 228, 231
Schäfer, Mirko, 62, 63
Scheidt, Lois A., 101
Schell, Eileen E., 11, 236
Schilb, John, 36
Schlosser, Eric, 30
Scott, J. Blake, 134
Scott, Tony, 22
Seattle (WA) and the World Trade Organization, 21
Second language acquisition (SLA), 139
Selber, Stuart, 180
Self, independent/interdependent, 160-164, 170-172
Selfe, Cynthia, 24, 49, 50, 188
Selfe, Richard, 49, 186, 188
Selvaratnam, Tanya, 83, 84
Selwyn, Neil, 184-185
Shaffer, Jan, 62
Shenhar, Aaron, 201
Shirky, Clay, 21-22
Sierra Club, 31
Simulation training. *See* Broadband divide and technical communication pedagogy
Single-source documentation. *See* XML documentation, rhetorical interventions for
Sites, Kevin, 71
Slatin, John, 43
Sobchack, Vivian, 90
Social media/relationships, 183
See also Friends offline, cultural/individual differences and meeting online
Socially responsible investing, 8
"Socrates in the Labyrinth," 51
Soundararajan, Tehnmozhi, 22
South Korea, 60, 65
See also Friends offline, cultural/individual differences and meeting online
South-North collaboration between professional writing students
alterity and ethics in the context of globalization, 135
background, 137
course descriptions and participants, 138-139
divergence, learning to navigate digital, 145-150
expectations, student, 141-144
globalization, 133-135
idealistic tone, 151
methods, study, 139-140

[South-North collaboration between professional writing students]
networked/informed/critical citizenship, 151
objectives, 137-138
overview, 133
positive experience, students view collaboration as a, 151, 154
potentials of cross-cultural online collaborations, 136-137
recommendations to those engaging in collaborations across different contexts, 150, 152-153
religious differences, 143, 144
reservations, U.S. students', 144
stereotypes, shaking off cultural, 134
Southwood, Russell, 5
Space metaphor and textual restructuring as encouraged by digital forms, 44
Spinuzzi, Clay, 190
Spokesman-Review's Daily Briefing Weblog, 69
Stahl, Elmar, 44
Stakeholder-rights movements, 8
Stallman, Richard, 62-63
Standard Occupational Classification (SOC) system, 100, 101
Staples, Katherine, 204
Starke-Meyerring, Doreen, 6, 23
Stein, Gertrude. *See* Performance in the digital age
Steward, Sherry, 12, 90, 182, 237
Stotts, David, 42
Suchan, James, 52
Sulzberger, Arthur O., Jr., 59
Summary Statement on Multimodal Literacies, 1
Sun, Huatong, 7
Supply-side approach to the dissemination of digital technology, 2
Surma, Anne, 51
Symbolic profit, 98
Synchronous communication, 162

Tacit knowledge, 120, 123
Taylor, T. L., 6
Technical and professional communication educators
assessment efforts, 184-189
[Technical and professional communication educators]
curricula and pedagogy, assessing, 185-186
faculty, assessing (self-assessment), 189
instructional spaces, assessing, 188-189
know about technologies/practices, what do students need to, 181-184
meaningful use of information/ communication technologies, 180, 190-192
overview, 179
partnerships with industry, 191-192
programs, assessing, 184-185
summary/conclusions, 192
textbooks and instructional materials, assessing, 187-188
threatening influence, digital technology constructed as a, 180
Technical Communication (Burnett), 139
Technical Communication in the Twenty-First Century (Dobrin, Keller, & Weisser), 187-188
Technical Communication Quarterly, 134, 188
Technology-justice relation model, 23
Textbooks used in technical/professional communication courses, 187-188
Textual restructuring as encouraged by digital forms
authorial norms/authority, the challenge to, 46-52
comprehension, implications for, 45-46
contradictions in print-to-digital and potential to energize composition, 40-41
dispositio (arrangement of that which has parts), contestation of, 35, 39, 41
fragmented argumentative arrangements, 37
overview, 35-36
reactions to print-to-digital, disciplinary, 41-44
resistance to divergence, 36-38
summary/conclusions, 52
Writing for the Information Age, 38-41
See also Performance in the digital age
Texuality, five global flows of, 139-140
Thailand, 163

Theater, live. *See* Performance in the digital age
Third World Majority (TWM), 22, 23
Torvalds, Linus, 57, 62, 66, 71, 73
"TPC Program Snapshots: Developing Curricula and Addressing Challenges" (Allen & Benninghoff), 180
Transcriptions Project, 191
Transparency issues and citizen journalism, 69-70
Tranter, Bruce, 5
Triandis, Harry, 159
Tripod, 103
Tunisia. *See* South-North collaboration between professional writing students
Turner, Roy, 182
Turnley, Melinda, 9, 23

United Nations, 4, 136
University of California at Santa Barbara, 191
U.S. National Telecommunications and Information Administration, 95
Users/designers binary, 6-7

Valk, Kate, 82, 84
See also Performance in the digital age
van den Boomen, Marianne, 62, 63
Van Dijk, Jan, 95
Vanishing Newspaper, The (Meyer), 59
Varis, Tapio, 163
Viral Flash activism. *See Meatrix, The*
Viral Spiral (Bollier), 26
Visibility of system status (VSS), 228
von Goethe, Johann W., 81
von Hippel, Eric, 66, 71-72

W. W. Norton, 38
Waldman, Simon, 59
Walters, Christian, 217
Walther, Joseph, 164
Warnick, Barbara, 37
Warschauer, Mark, 1
Washington Post, 68
Washington State Library's *Digital Best Practices* site, 10
WebCT, 9, 139, 140
Web 2.0 protocols, 46, 182, 183
Weiser, Irwin, 119
Weisser, Christian, 187
Welch, Kathleen, 188
We Media, 62
We the Media: Grassroots Journalism for the People by the People (Gillmor), 60
Whalen, Brandon, 1
"What Do We Need to Teach About Knowledge Work?" (Spinuzzi), 190
Where Where There There Where. *See* Performance in the digital age
Wick, Corey, 209
WIDE Research Center Collective, 188
Wilhelm, Anthony, 22
Williams, Sean, 188
Willis, Suzanne, 5
Winner, Langdon, 7
Wire and Plastics Products (WPP), 36-37
Wisconsin State Journal, 69
"Wither the Essay," 49, 52
Witt, Leonard, 11, 237
Wooster Group, 80-81
See also Performance in the digital age
World Trade Organization (WTO), 21
WPA. *See* Writing Program Administrators electronic mailing list
WPA Journal, 118
Wright, Elijah, 101
Writing across the curriculum (WAC), 127-128
Writing for the Information Age (Ross-Larson), 38-41
Writing in the disciplines (WID), 127-128
Writing Program Administrators electronic mailing list
circulation of existing formal knowledge, 121
collaboration and timely sharing of research finding, 123
community outreach, 126-128
compositionists in the conversation, involving, 126-127
creating/storing cultural knowledge about WPA as a profession, 125-126
emerging/new research and collaborative brainstorming, 122-123
formal *vs.* informal knowledge, 119-120
history of the WPA, 118-119
Internet/communication/writing, 117-118

[Writing Program Administrators electronic mailing list]
know-how or personal experience, 122
knowledge-making process, 120-121
overview, 117
participatory discourses, platform for, 123-124
practical know-how, collaborative creation of, 125
summary/conclusions, 128-129
tacit knowledge, articulation and circulation of, 123
training ground for newcomers to the field, 124-125
two-year colleges, involving teachers in, 127
writing across the curriculum and writing in the disciplines, 127-128
Writing Space (Bolter), 42

Xanadu, Project, 47, 48
XML documentation, rhetorical interventions for
advantages of XML, 214
audience and genre, overlooking dictates of, 217-218

[XML documentation, rhetorical interventions for]
Brio, 221-225
division of labor and interests in single-source documentation, 225-227
document type definition, 226, 227, 230
flexibility of XML syntax, 215-217
human-computer interaction and usability testing, 227-230
iCalibre, 218-221, 229
industry and the academy, 198
interoperability, 226
literate programming, 227
modular writing, 225-226
overview, 213-214
summary/conclusions, 230-231

Yahoo!, 71, 164

Zappen, James P., 37
Zuboff, Shoshana, 10